MEP 805B / 815B
Diesel Engine Parts Manual
TM 9-2815-259-24P

Diesel Engine
Model 4045TF151
4 Cylinder 4.5 Liter

edited by
Brian Greul

The MEP series of Military Generators are rugged, durable and incorporate proven diesel engine technology. This book is the diesel engine parts manual and also incorporates general and direct support instructions. It is being republished to assist enthusiasts, restorers, and aftermarket owners who use or wish to use these generators outside of military use.

An 8.5x11 3 hole punched loose leaf copy may be purchased for your 3 ring binder. Email books@ocotillopress.com for current information.

Should you have suggestions or feedback on ways to improve this book please send email to Books@OcotilloPress.com

Edited 2021 Ocotillo Press
ISBN 978-1-954285-21-7

Printed in the United States of America

Ocotillo Press
Houston, TX 77017
Books@OcotilloPress.com

Disclaimer: The user of this book is responsible for following safe and lawful practices at all times. The publisher assumes no responsibility for the use of the content of this book. The publisher has made an effort to ensure that the text is complete and properly typeset, however omissions, errors, and other issues may exist that the publisher is unaware of.

TECHNICAL MANUAL

UNIT, DIRECT SUPPORT AND GENER-
AL SUPPORT MAINTENANCE REPAIR PARTS
AND SPECIAL TOOLS LIST

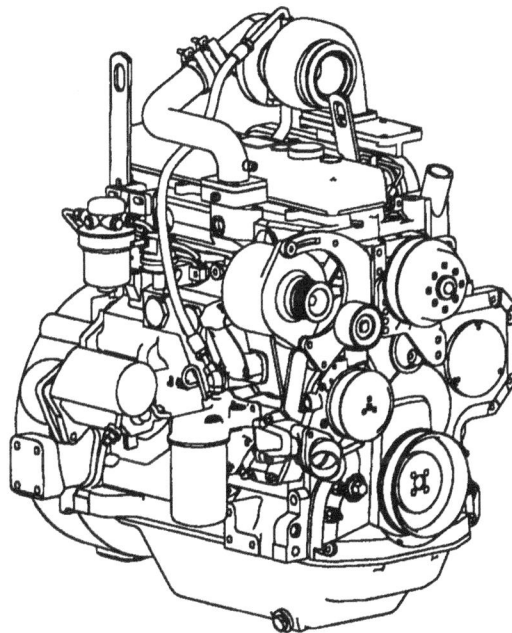

DIESEL ENGINE
MODEL 4045TF151
4 CYLINDER, 4.5 LITER
(NSN 2815-01-462-2289) (EIC: N/A)

DEPARTMENTS OF THE ARMY AND THE AIR FORCE
AND HEADQUARTERS, MARINE CORPS
1 NOVEMBER 2000

INSERT LATEST CHANGED PAGES, DESTROY SUPERSEDED PAGES

LIST OF EFFECTIVE PAGES

NOTE: The portion of the text affected by the changes is indicated by a vertical line in the outer margins of the page. Changes to illustrations are indicated by miniature pointing hands. Changes to wiring diagrams are indicated by shaded areas.

Dates of issue for original and changed pages are:

Original..........0.............1 Nov 2000

Page * Change Page * Change Page * Change No.   No. No. No. No. No.

Page	Change	Page	Change	Page	Change
Title	0	13-1	0	26-1	0
A/(B blank)	0	Fig 14 (1 sheet)	0	Fig 27 (1 sheet)	0
i-xi	0	14-1	0	27-1	0
Fig 1 (3 sheets)	0	Fig 15 (1 sheet)	0	Fig 28 (1 sheet)	0
1-1	0	15-1	0	28-1	0
Fig 2 (1 sheet)	0	Fig 16 (1 sheet)	0	Fig 29 (1 sheet)	0
2-1	0	16-1	0	29-1	0
Fig 3 (1 sheet)	0	Fig 17 (1 sheet)	0	Fig 30 (1 sheet)	0
3-1	0	17-1	0	30-1	0
Fig 4 (1 sheet)	0	Fig 18 (1 sheet)	0	Fig 31 (1 sheet)	0
4-1	0	18-1	0	31-1	0
Fig 5 (1 sheet)	0	Fig 19 (1 sheet)	0	Fig 32 (1 sheet)	0
5-1	0	19-1	0	32-1	0
Fig 6 (1 sheet)	0	Fig 20 (1 sheet)	0	Fig 33 (1 sheet)	0
6-1	0	20-1	0	33-1	0
Fig 7 (1 sheet)	0	Fig 21 (2 sheets)	0	Fig 34 (1 sheet)	0
7-1	0	21-1 - 21-3	0	34-1	0
Fig 8 (1 sheet)	0	Fig 22 (1 sheet)	0	Fig 35 (1 sheet)	0
8-1	0	22-1	0	35-1	0
Fig 9 (1 sheet)	0	Fig 23 (1 sheet)	0	Fig 36 (1 sheet)	0
9-1	0	23-1	0	36-1	0
Fig 10 (1 sheet)	0	Fig 24 (1 sheet)	0	Fig 37 (1 sheet)	0
10-1	0	24-1	0	37-1	0
Fig 11 (1 sheet)	0	Fig 25 (1 sheet)	0	Fig 38 (1 sheet)	0
11-1	0	25-1	0	38-1	0
Fig 12 (1 sheet)	0	Fig 26 (1 sheet)	0	I-1 – I-12	0
12-1	0				
Fig 13 (1 sheet)	0				

* Zero in this column indicates an original page.

ARMY TM 9-2815-259-24P
AIR FORCE TO 38G1-125-4
MARINE CORPS TM 09249A/2815-24P/4

Technical Manual
No. 9-2815-259-24P
Technical Order
No. 38G1-125-4
nical Manual
No. 09249A/2815-24P/4

HEADQUARTERS
DEPARTMENTS OF THE ARMY AND

THE AIR FORCE, AND

HEADQUARTERS, MARINE CORPS Tech-
WASHINGTON D.C., 1 NOVEMBER 2000

UNIT, DIRECT SUPPORT, AND GENERAL SUPPORT MAINTENANCE REPAIR PARTS AND SPECIAL TOOLS LIST

DIESEL ENGINE, MODEL 4045TF151
4 CYLINDER, 4.5 LITER

(NSN 2815-01-462-2289) (EIC: N/A)

REPORTING ERRORS AND RECOMMENDING IMPROVEMENTS

You can help improve this manual. If you find any mistakes or if you know of a way to improve the procedures, please let us know. Mail your letter, DA Form 2028 (Recommended Changes to Publications and Blank Forms), or DA 2028-2 located in the back of this manual, directly to: Commander, US Army Communications - Electronics Command and Fort Monmouth, ATTN: AMSEL-LC-LEO-D-CS-CFO, Fort Monmouth, New Jersey 07703-5006. The fax number is 732-532-1413, DSN 992-1413. You may also e-mail your recommendations to AMSEL-LC-LEO-PUBS-CHG@mail1.monmouth.army.mil

For Air Force, submit AFTO Form 22 (Technical Order System Publication Inprovement Report and Reply) In accordance with paragraph 6-5, Section VI, TO 00-5-1. Forward direct to prime ALC/MST.

Marine Corps units submit NAVMC 10772 (Recommended Changes to Technical Publications) to: Commanding General, Marine Corps Logistics Base (Code 850), Albany, Georgia 31704-5000.

A reply will be furnished to you.

TABLE OF CONTENTS

i

TABLE OF CONTENTS - continued

TABLE OF CONTENTS - continued

UNIT, DIRECT SUPPORT, AND
GENERAL SUPPORT MAINTENANCE
REPAIR PARTS AND SPECIAL TOOLS LIST

SECTION 1. INTRODUCTION

1. SCOPE.

This RPSTL lists and authorizes spares and repair parts; special tools; special test, measurement, and diagnostic equipment (TMDE); and other special support equipment required for performance of unit, direct support, and general support maintenance of the diesel engine. It authorizes the requisitioning, issue, and disposition of spares, repair parts, and special tools as indicated by the source, maintenance and recoverability (SMR) codes.

2. GENERAL.

In addition to Section I, Introduction, this Repair Parts and Special Tools List is divided into the following sections:

    a. Section II. Repair Parts List. A list of spares and repair parts authorized by this RPSTL for use in the performance of maintenance. The list also includes parts which must be removed for replacement of the authorized parts. Parts lists are composed of functional groups in ascending figure and item number sequence. Bulk materials are listed in item name sequence. Repair parts kits are listed separately in their own functional group within Section II. Repair parts for repairable special tools are also listed in this section. Items listed are shown on the associated illustration(s)/figure(s).

    b. Section III. Special Tools List. A list of special tools, special TMDE, and other special support equipment authorized by this RPSTL (as indicated by Basis of Issue (BOI) information in DESCRIPTION AND USABLE ON CODE column) for the performance of maintenance.

    c. Section IV. National Stock Number and Part Number Index. A list, in National Item Identification Number (NIIN) sequence, of all national stock numbered items appearing in the listing, followed by a list in alphanumeric sequence of all part numbers appearing in the listings. National stock numbers and part numbers are cross-referenced to each illustration figure and item number appearance.

3. EXPLANATION OF COLUMNS (SECTIONS II AND III).

    a. ITEM NO, (Column - (1)) . Indicates the number used to identify items called out in the illustration.

b. SMR CODE (Column (2)). The Source, Maintenance, and Recoverability (SMR) code is a 5-position code containing supply/requisitioning information, maintenance category authorization criteria, and disposition instruction, as shown in the following breakout:

```
       Source                    Maintenance      Recoverability   Code
       Code                      Code ☐ ☐_ ☐_ _____   1st two                    _____  _____  XX
       positions         XX              X  ☐                      ☐ ☐_
How you get an item  3rd position    4th position  Who determined  ☐ ☐                disposition ac-
                                     tion  Who can install,    Who can do  on an unserviceable  replace or use
                                     complete repair* item.   the item.    on the item.
```

* Complete Repair: Maintenance capacity, capability, and authority to perform all corrective maintenance tasks of the "Repair" function in a use/user environment in order to restore serviceability to a failed item.

(1) Source Code. The source code tells you how to get an item needed for maintenance, repair, or overhaul of an end item/equipment. Explanations of source codes follows:

Code   Explanation

PA
PB   Stocked items; use the applicable NSN to request/requisition items with these source codes.
PC**   They are authorized to the category indicated by the code entered in the 3rd position
PD   of the SMR code.
PE
PF
PG   ** NOTE: Items coded PC are subject to deterioration.

Code   Explanation

KD   Items with these codes are not to be requested/requisitioned individually. They are part of a KF kit which is authorized to the maintenance category indicated in the 3rd position of the SMR
KB   code. The complete kit must be requisitioned and applied.

Code   Explanation

MO - (Made at org or unit)   Items with these codes are not to be requested/requisitioned individ- MF - (Made at DS or intermediate) ually. They must be made from bulk material which is identified MH - (Made at GS) by the part number in the DESCRIPTION AND USABLE ON CODE ML - (Made at Specialized Repair (UOC) column and listed in the Bulk Material group of the repair Activity (SRA)) parts list in this RPSTL. If the item is authorized to you by the 3rd MD - (Made at Depot) 3rd position code of the SMR code, but the source code indicates it is

made at a higher category, order the item from the higher category.

Code   Explanation

AO - (Assembled by org or unit) Items with these codes are not to be requested/requisitioned AF - (Assembled by DS or individually. The parts that make up the assembled item must be

intermediate) requisitioned or fabricated and assembled at the category of AH - (Assembled by GS) maintenance indicated by the source code. If the 3rd position code of AL - (Assembled by SRA) the SMR authorizes you to replace the item, but the source code AD - (Assembled by Depot) indicates the item is assembled at a higher category, order the item

from the higher category of maintenance.

Code   Explanation

XA - Do not requisition an "XA"-coded item. order its next higher assembly. (Also, refer to the NOTE below.)

XB - If an "XB" item is not available from salvage, order it using the FSCM and part number given. XC - Installation drawing, diagram, instruction sheet, field service drawing, that is identified by manufacturer's part number.

XD - Item is not stocked. Order an "XD"-coded item through normal supply channels using the FSCM and part number given, if no NSN is available.

NOTE: Cannibalization or controlled exchange, when authorized, may be used as a source of supply f or items with the above source codes, except for those source coded "XA" or those aircraft support items restricted by requirements of AR 700-42.

(2) Maintenance Code. Maintenance codes tell you the category (s) of maintenance authorized to USE and REPAIR support items. The maintenance codes are entered in the 3rd and 4th positions of the SMR Code as follows:

(a) The maintenance code entered in the 3rd position tells you the lowest maintenance category authorized to remove, replace, and use an item. The maintenance code entered in the 3rd position will indicate authorization to one of the following categories of maintenance.

Code   Application/Explanation

C   Crew or operator maintenance done within organization or unit maintenance.

O   Organizational or unit category can remove, replace, and use the item.

F   Direct support or intermediate category can remove, replace, and use the item.

H   General support catagory can remove, replace, and use the item.

L   Specialized repair activity can remove, replace, and use the item.

D   Depot catagory can remove, replace, and use the item.

(b) The maintenance code entered in the 4th position tells whether or not the item is to be repaired and identifies the lowest maintenance category with the capability to do complete repair (i.e., perform all authorized repair functions).

> NOTE: Some limited repair may be done on the item at a lower category of maintenance, if authorized by the Maintenance Allocation Chart (MAC) and SMR Codes). This position will contain one of the following maintenance codes.

Maintenance
Code   Application/Explanation

O   Organizational or unit is the lowest category that can do complete repair of the item.

F   Direct support or intermediate is the lowest category that can do complete repair of the item.

H   General support is the lowest category that can do complete repair of the item.

L   Specialized repair activity (designate the specialized repair activity) is the lowest category that can do complete repair of the item.

D   Depot is the lowest category that can do complete repair of the item.

Z   Nonrepairable. No repair is authorized.

B   No repair is authorized. (No parts or special tools are authorized for the maintenance of a "B" coded item). However, the item may be reconditionedbyadjusting, lubricating, etc., at the user category.

(3) Recoverability Code. Recoverability codes are assigned to items to indicate the disposition action on unserviceable items. The recoverability code is entered in the 5th position of the SMR Code as follows:

Recoverability
Code   Application/Explanation

Z   Nonrepairable item. When unserviceable, condemn and dispose of the item at the category of maintenance shown in 3rd position of SMR Code.

O   Repairable item. When uneconomically repairable, condemn and dispose of the item at organizational or unit category.

F   Repairable item. when uneconomically repairable, condemn and dispose of the item at the direct support or intermediate category.

H   Repairable item. when uneconomically repairable, condemn and dispose of the item at the general support category.

Recoverability
Code   Application/Explanation

D   Repairable item. When beyond lower category repair capability, return to depot. Condemnation and disposal of item not authorized below depot category.

L   Repairable item. Condemnation and disposal not authorized below specialized repair activity (SRA).

A   Item requires special handling or condemnation procedures because of specific reasons (e.g., precious metal content, high dollar value, critical material, or hazardous material). Refer to appropriate manuals/directives for specific instructions.

c. CAGEC (Column (3)). The Commercial and Government Entity Code (CAGEC) is a 5-digit numeric code which is used to identifythemanufacturer, distributor, orgovernmentagency, etc., that supplies the item.

d. PART NUMBER (Column (4)). Indicates the primary number used by the manufacturer (individual, company" firm, corporation, or Government activity), which controls the design and characteristics of the item by means of its engineering drawings, specifications standards, and inspection requirements to identify an item or range of items.

NOTE: When you use a NSN to requisition an item, the item you receive may have a different part number from the part ordered.

e. DESCRIPTION AND USABLE ON CODE (UOC) (Column (5)). This column includes the following information:

`(1) The Federal item name and, when required, a minimum description to identify the item.

(2) Items that are included in kits and sets are listed below the name of the kit or set.

(3) Spare/repair parts that make up an assembled item are listed immediately following the assembled item line entry.

(4) Part numbers for bulk materials are referenced in this column in the line item entry for the item to be manufactured/fabricated.

(5) When the item is not used with all serial numbers of the same model, the effective serial numbers are shown on the last line(s) of the description (before UOC).

(6) The UOC, when applicable (see paragraph 5, Special Information).

(7) In the Special Tools List section, the Basis of Issue (BOI) appears as the last line(s) in the entry for each special tool, special TMDE, and other special support equipment. When density of equipments supported exceeds density spread indicated in the BOI, the total authorization is increased proportionately.

(8) The statement "END OF FIGURE" appears just below the last item description in Column 5 for a given figure in Section II.

f. OTY (Column (6)). The QTY (quantity per figure column) indicates the quantity of the item used in the breakout shown on the illustration figure, which is prepared for a functional group, subfunctional group, or an assembly. A "V" appearing in this column in lieu of a quantity indicates that the quantity is variable and that the quantity may vary from application to application.

g. EXPLANATION OF COLUMNS (SECTION IV).

a. National Stock Number (NSN) Index.

(1) STOCK NUMBER Column. This column lists the NSN by National Item Identification Number (NIIN) sequence. The NIIN consists of the last nine digits of the NSN.

$$\frac{\overline{\text{NSN}}}{(\ 5305 - \underline{01 - 674 - 1467}\ )}$$
$$\text{NIIN}$$

When using this column to locate an item, ignore the first 4 digits of the NSN. However, the complete NSN should be used when ordering items by stock number.

(2) FIG. Column. This column lists the number of the figure where the item is identified/located. The figures are in numerical order in Section II and Section III.

(3) ITEM Column. The item number identifies the item associated with the figure listed in the adjacent FIG. Column. This item is also identified by the NSN listed on the same line.

b. Part Number Index. Part numbers in this index are listed by part number in ascending alphanumeric sequence (i.e., vertical arrangement of letter and number combination which places the first letter or digit of each group in order A through Z, followed by numbers 0 through 9 and each following letter or digit in like order).

(1) CAGEC Column. The Commercial and Government Entity Code (CAGEC) is a 5-digit numeric code used to identify the manufacturer, distributor, or Government agency, etc., that supplies the item.

(2) PART NUMBER Column. Indicates the primary number used by the manufacturer (individual, firm, corporation, or Government activity) , which controls the design and characteristics of the item by means of its engineering drawings, specifications standards, and inspection requirements to identify an item or range of items.

(3) STOCK NUMBER Column. This column lists the NSN for the associated part number and manufacturer identified in the PART NUMBER and FSCM columns to the left.

(4) FIG. Column. This column lists the number of the figure where the item is identified/located in Section II and III.

(5) ITEM Column. The item number is that number assigned to the item as it appears in the figure referenced in the adjacent figure number column.

5. SPECIAL INFORMATION. Use the following subparagraphs as applicable:

6. <u>Usable on Code.</u> <u>The usable on code appears in the lower left corner of the Description column heading.</u> Usable on codes are shown as "UOC: . . ." under the applicable item description/nomenclature. Uncoded items are applicable to all models. Identification of the usable on codes used in the RPSTL are:

<u>Code</u>	Used On
LTZ	Diesel Engine, 4 Cylinder, Model 4045TF151

b. <u>Fabrication Instructions.</u> <u>Bulk materials required to manufacture items are listed in the Bulk Material</u> Functional Group of this RPSTL. Part numbers for bulk materials are also referenced in the description column of the line item entry for the item to be manufactured/fabricated. Detailed fabrication instructions for item source codes to be manufactured or fabricated are found in TM 9-2815-259-24.

c. <u>Assembly Instruction.</u> <u>Detailed assembly instructions f or items source coded to be assembled from</u> component spare/repair parts are found in TM 9-2815-259-24. Items that make up the assembly are listed immediately following the assembly item entry or reference is made to an applicable figure.

d. <u>Kits.</u> Line item entries for repair parts kits appear throughout Section II.

e. <u>Associated Publications.</u> <u>The publication(s) listed below pertain to this engine its components:</u>

<u>Publication</u>	<u>Short Title</u>
TM 9-2815-259-24	Maintenance Manual, Diesel Engine, Model 4045TF151

f. <u>National Stock Numbers.</u> <u>National stock numbers (NSN's) that are missing from "P" source coded</u> <u>items</u> have been applied for and will be added to this TM by future change/revision when they are entered in the Army Master Data File (AMDF). Until the NSN's are established and published, submit exception requisitions to:
Commander, US Army Communications-Electronics Command and Fort Monmouth, ATTN: AMSEL-LC-MM, Fort Monmouth, NJ 07703-5007 for the part required to support your equipment.
6. HOW TO LOCATE REPAIR PARTS.

a. <u>When National Stock Number or Part Number is Not Known.</u>

(1) <u>First</u>. Using the Table of Contents, determine the assembly group or subassembly group to which the item belongs. This is necessary since figures are prepared for assembly groups and subassembly groups, and listings are divided into the same groups.

(2) <u>Second</u>. Find the figure covering the assembly group or subassembly group to which the item belongs.

(3) <u>Third</u>. Identify the item on the figure and note the item number.

(4) <u>Fourth.</u> Refer to the Repair Parts List for the figure to find the part number for the item number noted on the figure.

(5) <u>Fifth.</u> Refer to the Part Number Index to find the NSN, if assigned.

b. <u>When National Stock Number or Part Number is Known.</u>

(1) <u>First.</u> Using the Index of National Stock Numbers and Part Numbers, find the pertinent National Stock Number or Part Number. The NSN index is in National Item Identification Number (NIIN) sequence (see 4. a (1)) . The part numbers in the Part Number index are listed in ascending alphanumeric sequence (see 4.b). Both indexes cross-reference you to the illustration figure and item number of the item you are looking for.

(2) <u>Second.</u> After finding the figure and item number, verify that the item is the one you're looking for, then locate the item number in the repair parts list for the figure.

7. ABBREVIATIONS.

<u>Abbreviations</u> <u>Explanation</u> BOI   Basis of Issue

DS   Direct Support

GS   General Support

MAC   Maintenance Allocation Chart

NIIN   National Item Identification Number
(consists of the last 9-digits of the NSN)

NSN   National Stock Number

RPSTL   Repair Parts and Special Tools List

SMR   Source, Maintenance and Recoverability Codes SRA   Special Repair Activity

TMDE   Test, Measurement and Diagnostic Equipment UOC   Usable on Code

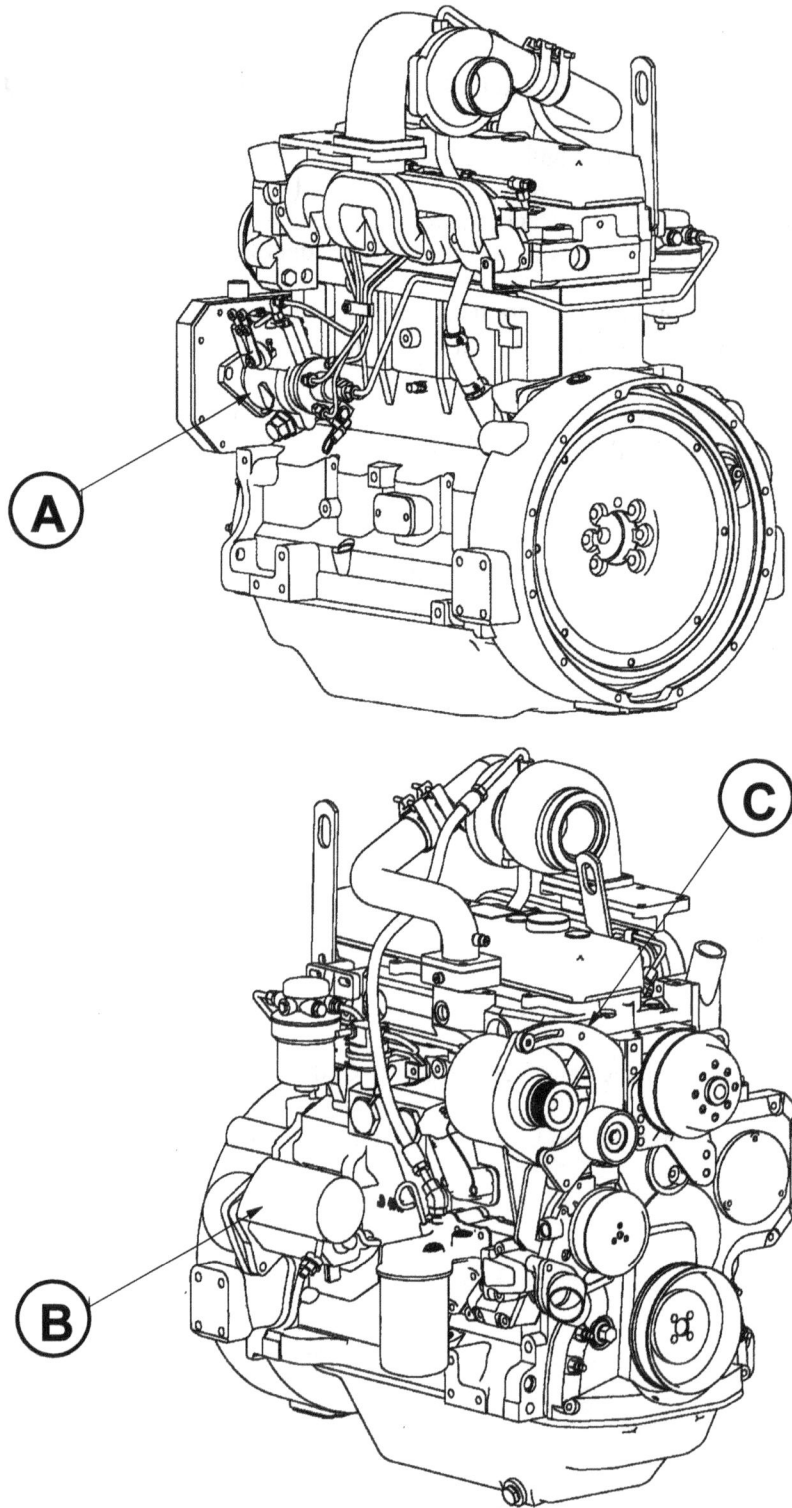

Figure 1. Engine Assembly (Sheet 1 of 3)

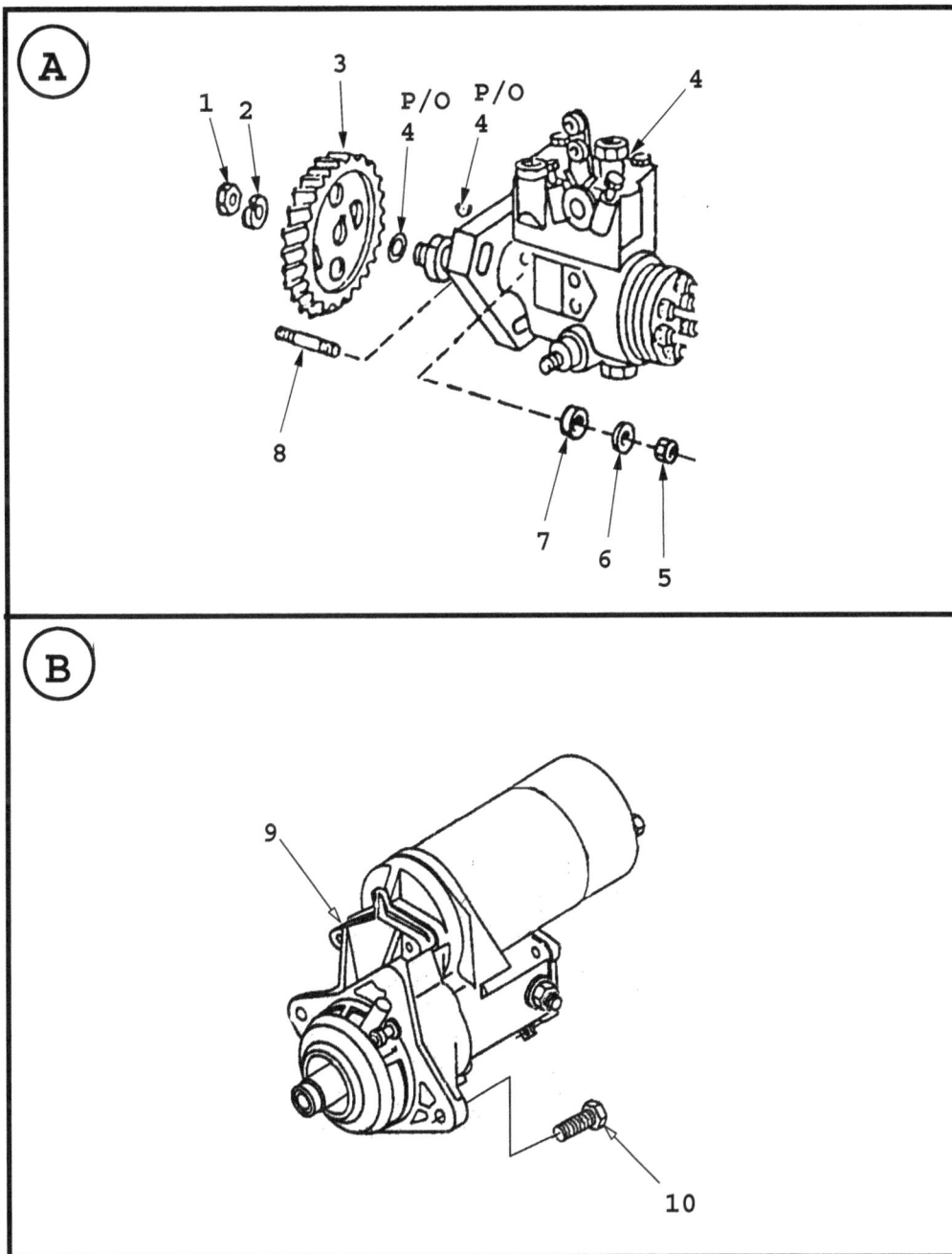

Figure 1. Engine Assembly (Sheet 2 of 3)

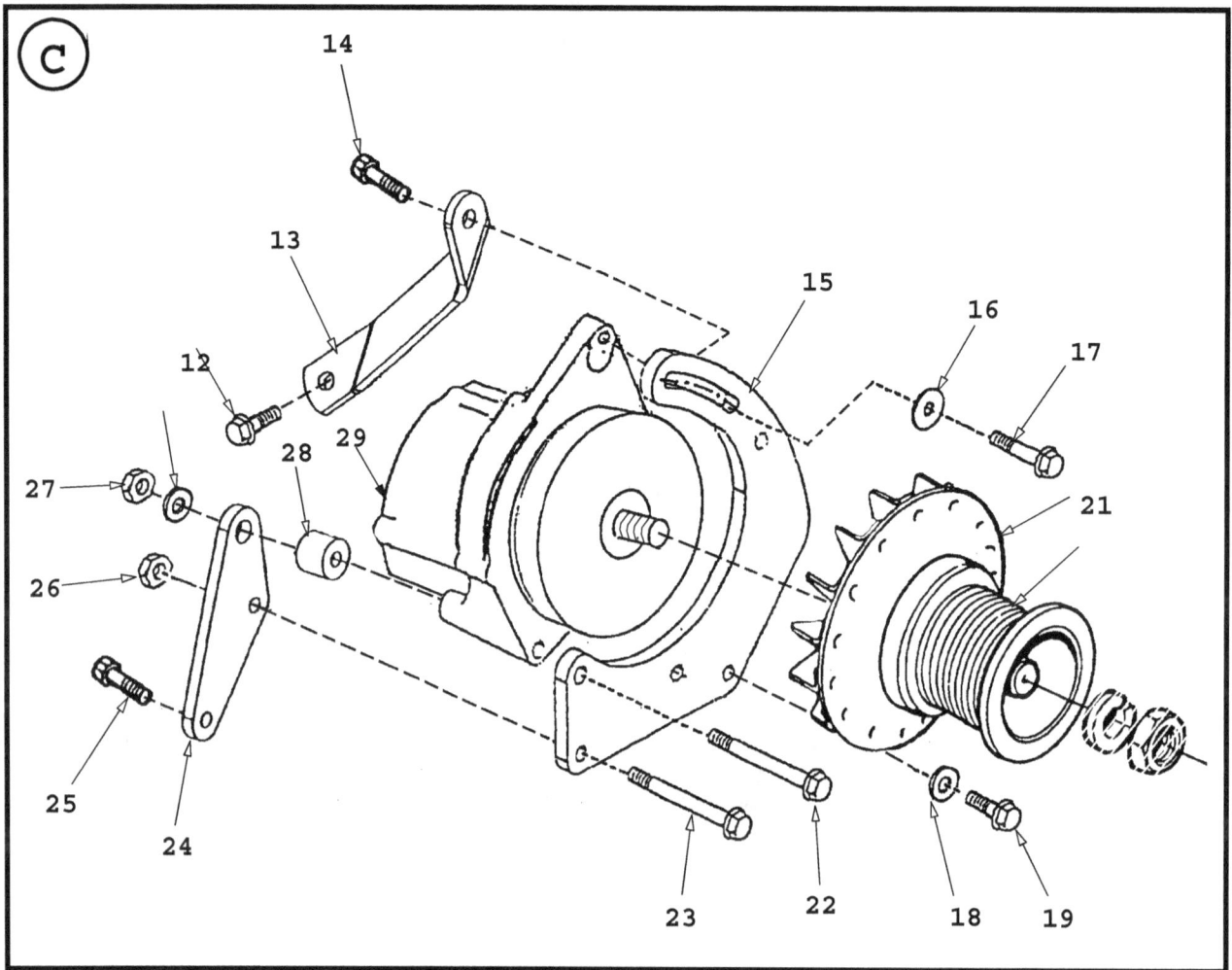

Figure 1. Engine Assembly (Sheet 3 of 3)

(1)			(3)	(4)	(5)	(6)
SMR CODE	(2)				DESCRIPTION AND	
ITEM	AIR			PART		
NO ARMY	FORCE	USMC	CAGEC	NUMBER	USABLE ON CODE (UOC)	QTY

GROUP 00 ENGINE ASSEMBLY
FIGURE 1 ENGINE ASSEMBLY

1	PAFZZ	PAFZZ	PAFZZ	75160	R91360	NUT,PLAIN,HEXAGON....................1
2	PAFZZ	PAFZZ	PAFZZ	66836	R132874	WASHER..............................1
3	XBFZZ	XB	XBFZZ	66836	R132267	GEAR,SPUR...........................1 4
	PAFHH	PAFFF	PAFHH	66836	RE67563	PUMP,INJECTION,FUEL
						(REFER TO FIG. 21 FOR BREAKDOWN).....1
5	PAFZZ	PAFZZ	PAFZZ	66836	14M7273	NUT,PLAIN,HEXAGON...................3
6	PAFZZ	PAFZZ	PAFZZ	75160	12M7065	WASHER,FLAT.........................3
7	PAFZZ	PAFZZ	PAFZZ	75755	T27860	WASHER,FLAT.........................3
8	PAHZZ	PAFZZ	PAHZZ	66836	R116386	STUD,PLAIN.......................... 3 9
	PAOFF	PAFFF	PAOFF	75755	RE48827	STARTER,ENGINE,ELEC
						(REFER TO FIG. 5 FOR BREAKDOWN)......1
10	PAOZZ	PAOZZ	PAOZZ	75755	19M7786	BOLT,MACHINE........................2
11	PAOZZ	PAOZZ	PAOZZ	66836	R64525	WASHER,FLAT.........................1
12	PAOZZ	PAOZZ	PAOZZ	75755	19M7785	BOLT,MACHINE........................1
13	XBOZZ	XB	XBOZZ	66836	R500333	SUPPORT,BRACKET.....................1
14	PAOZZ	PAOZZ	PAOZZ	75755	19M7783	BOLT,MACHINE........................1
15	XBOZZ	XB	XBOZZ	66836	R500079	ARM,ADJUSTING,BELT..................1
16	PAOZZ	PAOZZ	PAOZZ	75160	R48808	WASHER,FLAT.........................1
17	PAOZZ	PAOZZ	PAOZZ	75755	19M7867	BOLT,MACHINE........................1
18	PAOZZ	PAOZZ	PAOZZ	66836	24M7207	WASHER..............................1
19	PAOZZ	PAOZZ	PAOZZ	66836	19M8039	SCREW,FLANGED.......................1
20	XBOZZ	XB	XBOZZ	66836	R94350	PULLEY..............................1
21	XBOZZ	XB	XBOZZ	66836	T158584	FAN,CENTRIFUGAL.....................1
22	PAOZZ	PAOZZ	PAOZZ	66836	19M7812	BOLT,MACHINE........................1
23	PAOZZ	PAOZZ	PAOZZ	66836	19M7913	SCREW,FLANGED.......................1
24	XBOZZ	XB	XBOZZ	66836	R500080	BRACKET,MOUNTING....................1
25	PAOZZ	PAOZZ	PAOZZ	66836	19M7868	SCREW,SERRATED,FLAN.................1
26	PAOZZ	PAOZZ	PAOZZ	75755	14M7298	NUT,HEX,FLANGED.....................1
27	PAOZZ	PAOZZ	PAOZZ	66836	14M7296	NUT,HEX,FLANGED.....................1
28	PAOZZ	PAOZZ	PAOZZ	66836	R500724	SPACER..............................1 29
	PAOFF	PAOFF	PAOFF	66836	AT175195	ALTERNATOR,24V,45A
						(REFER TO FIG. 4 FOR BREAKDOWN)......1

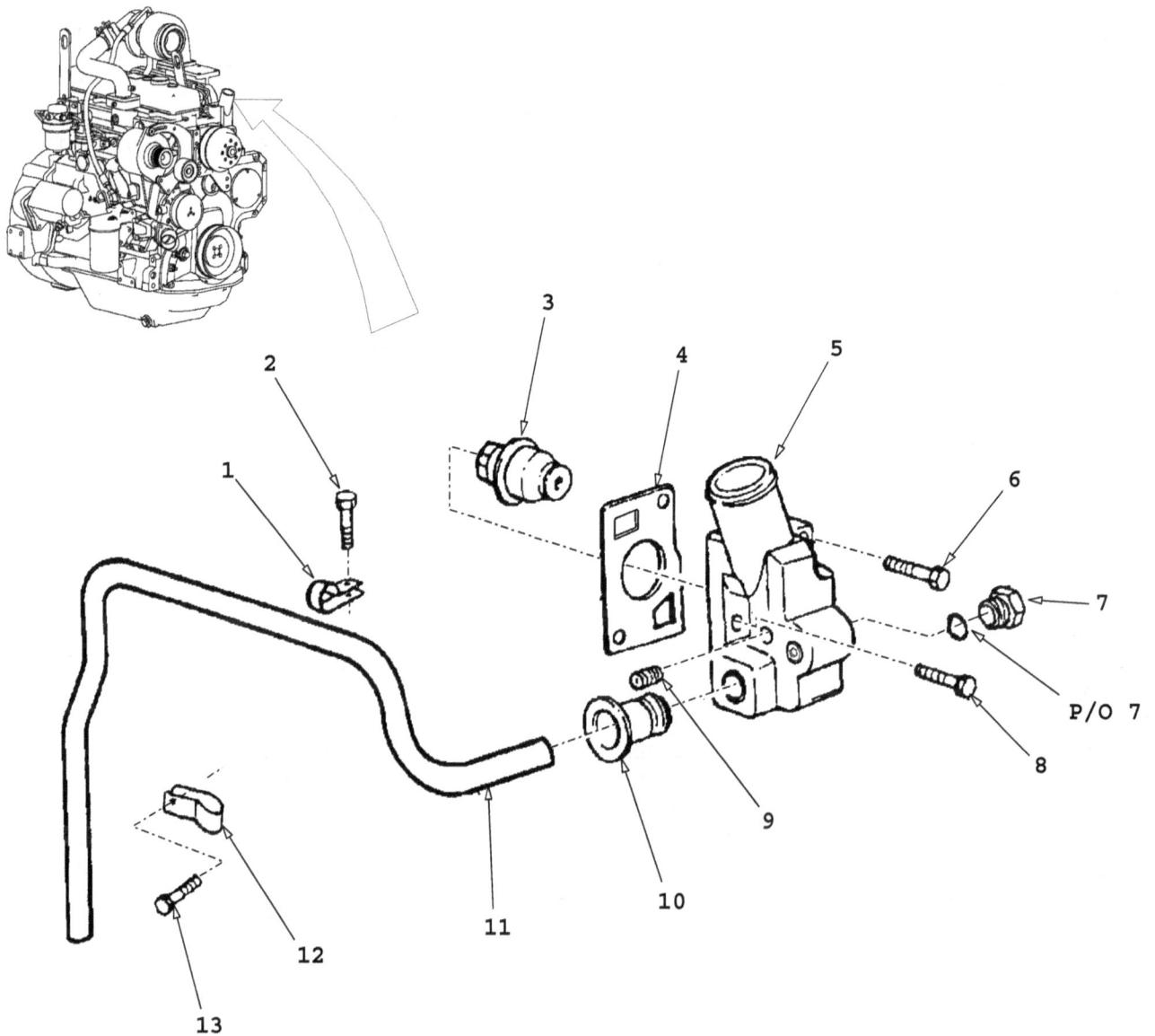

Figure 2. Thermostat and Water Manifold/Thermostat Cover

(1)				(3)	(4)	(5)	(6)
SMR CODE							
ITEM		AIR			PART	DESCRIPTION AND	
NO	ARMY	FORCE	USMC	CAGEC	NUMBER	USABLE ON CODE (UOC)	QTY

GROUP 0101 THERMOSTAT AND WATER MANIFOLD/THERMOSTAT COVER
FIGURE 2 THERMOSTAT AND WATER MANIFOLD/THERMOSTAT COVER

1	PAOZZ	PAOZZ	PAOZZ	66836	R123323	CLAMP,HOSE...........................1
2	PAOZZ	PAOZZ	PAOZZ	66836	19M7788	BOLT,MACHINE........................2
3	PAOZZ	PAOZZ	PAOZZ	66836	R135896	GASKET...............................1
4	PAOZZ	PAOZZ	PAOZZ	75160	RE64354	THERMOSTAT,FLOW CON.................1
5	XBOZZ	XB	XBOZZ	66836	R123426	COVER,THERMOSTAT....................1
6	PAOZZ	PAOZZ	PAOZZ	66836	19M7809	BOLT,MACHINE........................1
7	XBOZZ	XB	XBOZZ	66836	RE46684	PLUG................................1
8	PAOZZ	PAOZZ	PAOZZ	75755	19M7786	BOLT,MACHINE........................1
9	XBOZZ	XB	XBOZZ	75160	15H697	FITTING,PIPE PLUG...................3
10	PAOZZ	PAOZZ	PAOZZ	66836	R123226	SEAL,TUBE...........................1
11	XBOZZ	XB	XBOZZ	66836	R501029	TUBE,METALLIC,BYPASS................1
12	PAOZZ	PAOZZ	PAOZZ	66836	R123324	CLAMP,HOSE 1
13	PAOZZ	PAOZZ	PAOZZ	66836	19M7801	SCREW...............................1

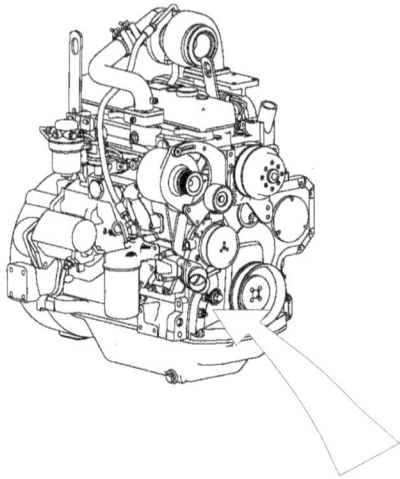

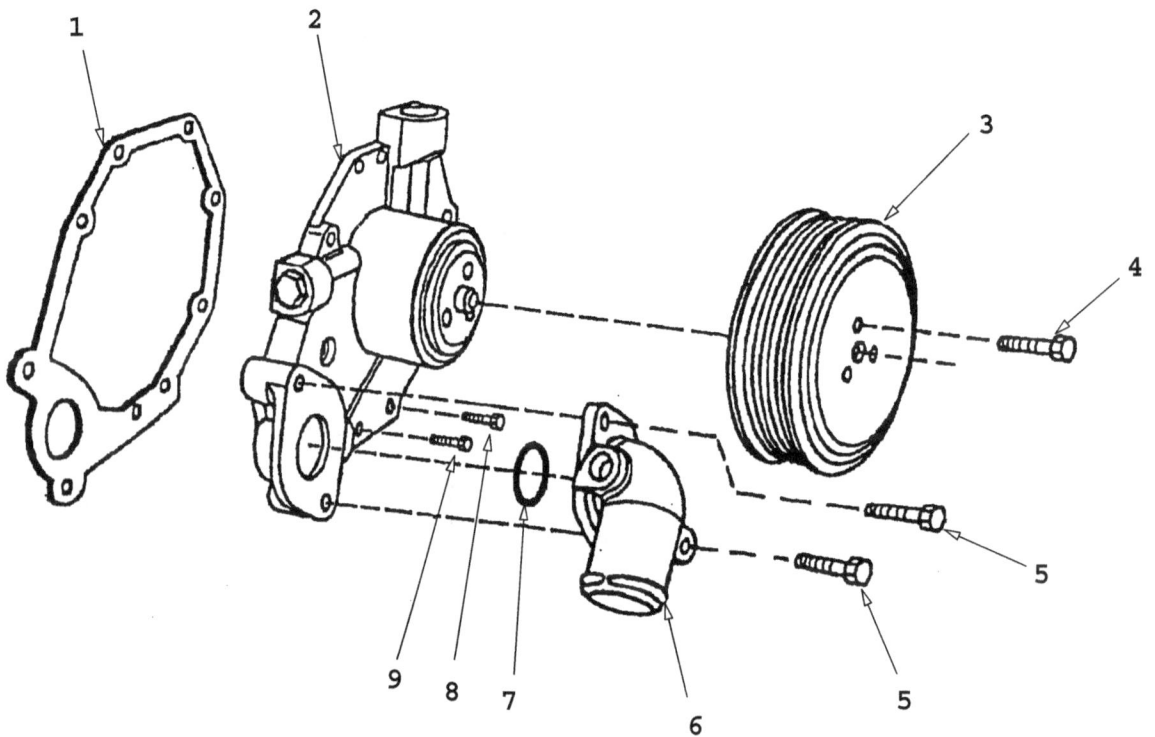

Figure 3. Water Pump

(1) SMR CODE ITEM NO	(2) ARMY	AIR FORCE	USMC	(3) CAGEC	(4) PART NUMBER	(5) DESCRIPTION AND USABLE ON CODE (UOC)	(6) QTY
						GROUP 0102 WATER PUMP FIGURE 3 WATER PUMP	
1	PAOZZ	PAOZZ	PAOZZ	75160	R123417	GASKET................................1	
2	PAOZZ	PAOZZ	PAOZZ	66836	RE500737	PUMP,COOLING SYSTEM,WATER............1	
3	XBOZZ	XB	XBOZZ	66836	R121038	PULLEY,WATER PUMP....................1	
4	PAOZZ	PAOZZ	PAOZZ	75755	19M7774	BOLT,MACHINE.........................3	
5	PAOZZ	PAOZZ	PAOZZ	75755	19M7867	BOLT,MACHINE.........................2	
6	XBOZZ	XB	XBOZZ	66836	RE501453	ELBOW,TUBE...........................1	
7	PAOZZ	PAOZZ	PAOZZ	75160	R89944	O-RING...............................1	
8	PAOZZ	PAOZZ	PAOZZ	66836	19M7863	BOLT,MACHINE.........................5	
9	PAOZZ	PAOZZ	PAOZZ	66836	19M7796	BOLT,MACHINE.........................3	

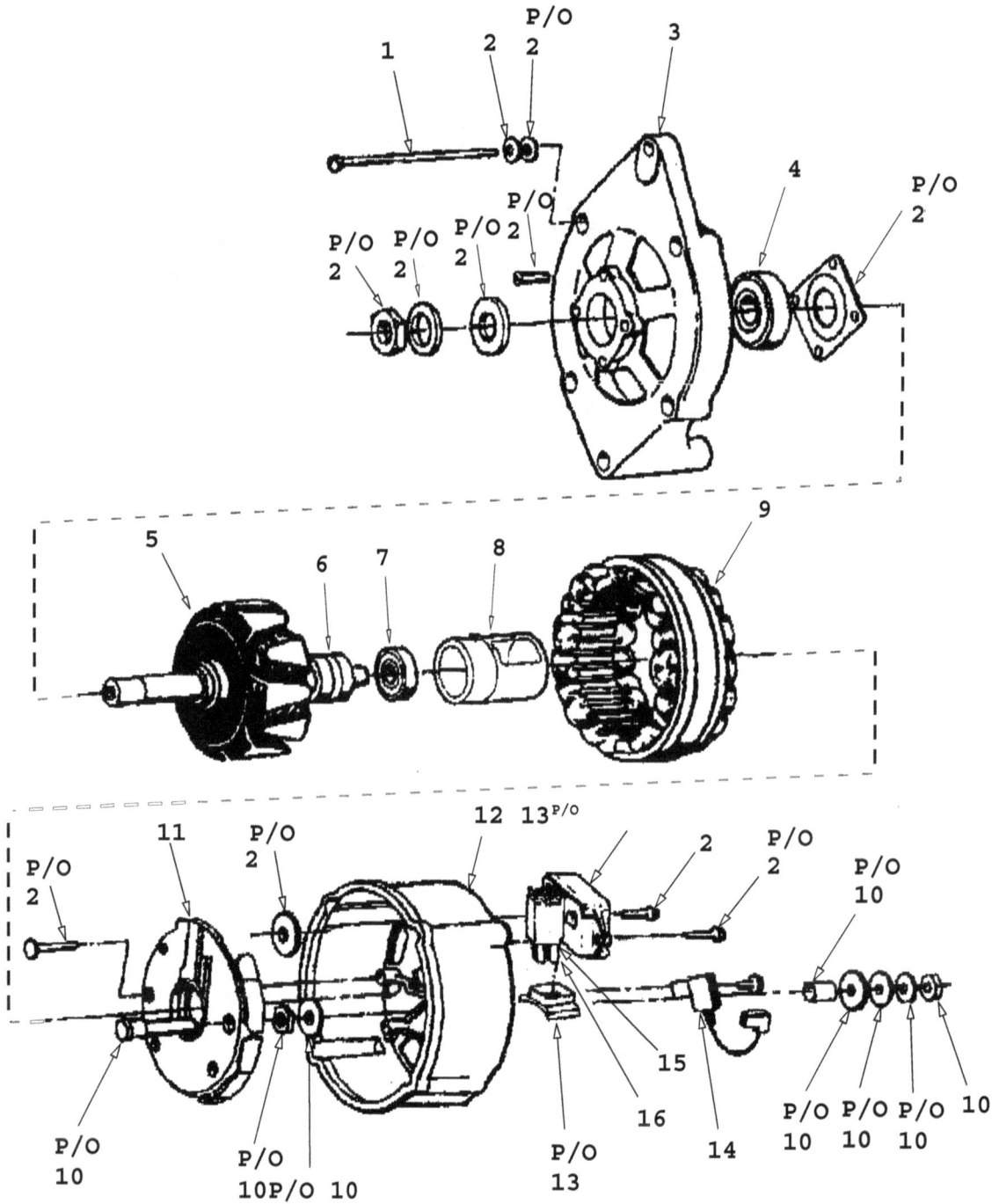

Figure 4. Battery Charging Alternator

(1) SMR CODE			(2)	(3)	(4)	(5)	(6)
ITEM NO	ARMY	AIR FORCE	USMC	CAGEC	PART NUMBER	DESCRIPTION AND USABLE ON CODE (UOC)	QTY

GROUP 0201 BATTERY CHARGING ALTERNATOR
FIGURE 4 BATTERY CHARGING ALTERNATOR

1	PAFZZ	PAFZZ	PAFZZ	53867	1123429200	BOLT,HEX HEAD	4
2	PAFZZ	PAFZZ	PAFZZ	53867	9122067056	PARTS KIT,REPAIR	1
3	XBFZZ	XB	XBFZZ	53867	9123065832	HOUSING,DRIVE END	1
4	PAFZZ	PAFZZ	PAFZZ	53867	9900069006	BEARING,BALL	1
5	PAFZZ	PAFZZ	PAFZZ	53867	9121064071	ROTOR ASSEMBLY	1
6	PAFZZ	PAFZZ	PAFZZ	53867	1124303024	SERVICE KIT,SLIPRING	1
7	PAFZZ	PAFZZ	PAFZZ	53867	9900069008	BEARING,BALL	1
8	XBFZZ	XB	XBFZZ	53867	1120591075	PLATE,METAL	1
9	PAFZZ	PAFZZ	PAFZZ	53867	9121064255	STATOR	1
10	PAFZZ	PAFZZ	PAFZZ	53867	9122067026	SERVICE KIT,TERMINA	1
11	PAFZZ	PAFZZ	PAFZZ	53867	9122067305	RECTIFIER	1
12	XBFZZ	XB	XBFZZ	53867	9123065806	HOUSING,SLIPRING EN	1
13	PAFZZ	PAFZZ	PAFZZ	53867	9191067042	REGULATOR ASSEMBLY	1
14	PAFZZ	PAFZZ	PAFZZ	53867	0290800069	CAPACITOR	1
15	PAFZZ	PAFZZ	PAFZZ	53867	9121067023	SERVICE KIT,SPRING	1
16	PAFZZ	PAFZZ	PAFZZ	53867	GB610	SERVICE KIT,BRUSH	1

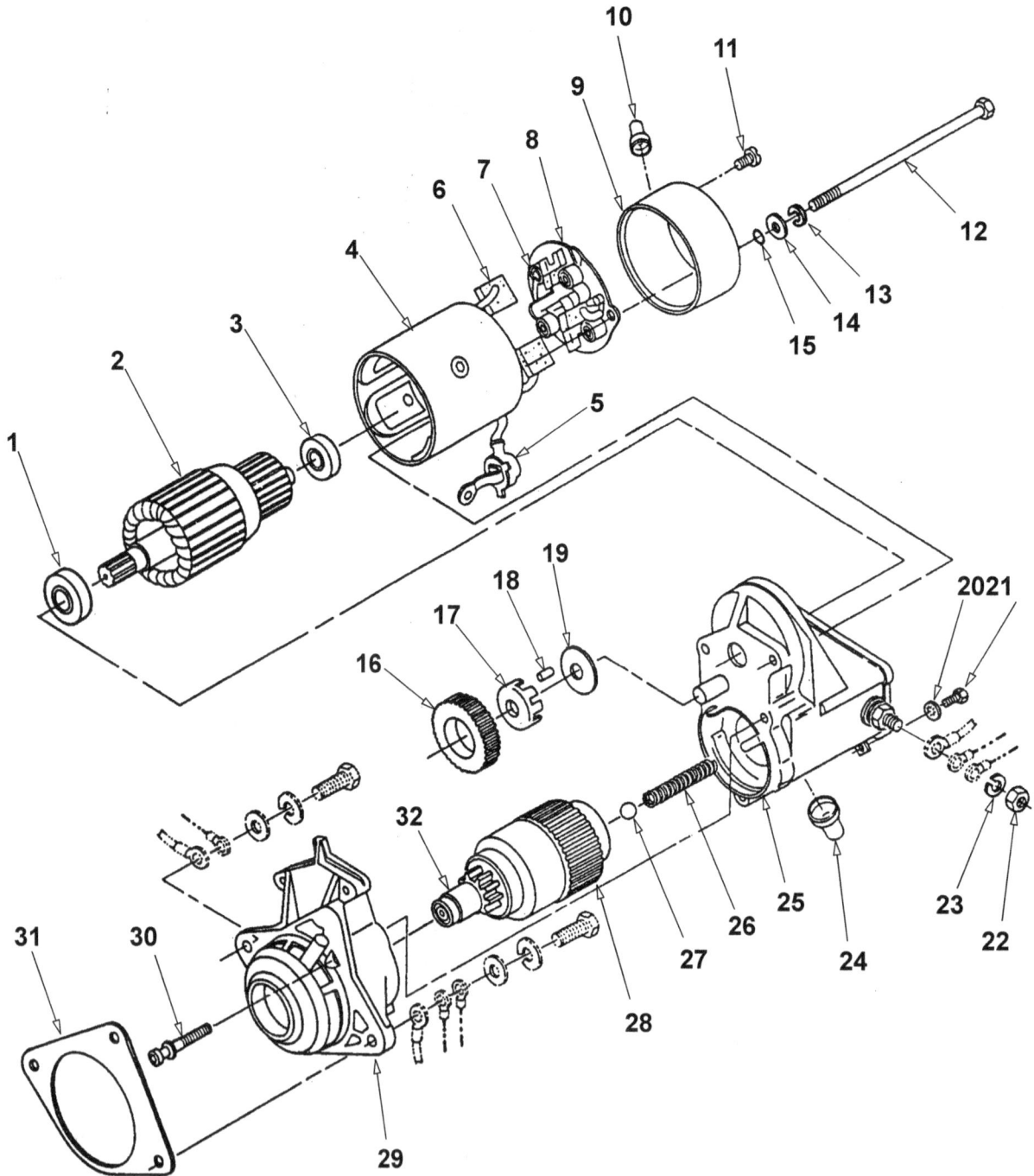

Figure 5. Starter

(1) SMR CODE ITEM NO ARMY	(2) AIR FORCE	USMC	(3) CAGEC	(4) PART NUMBER	(5) DESCRIPTION AND USABLE ON CODE (UOC)	(6) QTY
					GROUP 0202 STARTER FIGURE 5 STARTER	
1 PAFZZ	PAFZZ	PAFZZ	SA352	949100-2210	BEARING,BALL...........................1	
2 PAFZZ	PAFZZ	PAFZZ	SA352	028200-2230	ARMATURE ASSEMBLY....................1	
3 PAFZZ	PAFZZ	PAFZZ	SA352	949100-2160	BEARING,BALL...........................1	
4 PAFZZ	PAFZZ	PAFZZ	SA352	028100-2881	YOKE ASSEMBLY.........................1	
5 XBFZZ	XB	XBFZZ	SA352	028062-0130	COVER....................................1	
6 PAFZZ	PAFZZ	PAFZZ	SA352	028099-3570	SERVICE KIT,BRUSH....................2	
7 PAFZZ	PAFZZ	PAFZZ	SA352	949173-0571	SPRING...................................4	
8 XBFZZ	XB	XBFZZ	SA352	028510-0500	HOLDER,BRUSH.........................1	
9 XBFZZ	XB	XBFZZ	SA352	028501-6900	FRAME....................................1	
10 XBFZZ	XB	XBFZZ	SA352	028062-5790	COVER....................................1	
11 PAFZZ	PAFZZ	PAFZZ	SA352	949007-6530	SCREW,FLANGED.......................2	
12 PAFZZ	PAFZZ	PAFZZ	SA352	949042-2270	BOLT,HEXAGON HEAD...................2	
13 PAFZZ	PAFZZ	PAFZZ	SA352	90258-06001	WASHER,LOCK...........................2	
14 PAFZZ	PAFZZ	PAFZZ	SA352	949011-5700	WASHER,FLAT...........................2	
15 PAFZZ	PAFZZ	PAFZZ	SA352	90804-10050	O-RING....................................2	
16 PAFZZ	PAFZZ	PAFZZ	SA352	028371-1500	PINION....................................1	
17 PAFZZ	PAFZZ	PAFZZ	SA352	028063-0030	RETAINER................................1	
18 XBFZZ	XB	XBFZZ	SA352	028306-0070	ROLLER,CLUTCH........................5	
19 PAFZZ	PAFZZ	PAFZZ	SA352	028327-0171	RETAINER,SNAP........................1	
20 PAFZZ	PAFZZ	PAFZZ	SA352	949016-0220	WASHER,LOCK...........................1	
21 PAFZZ	PAFZZ	PAFZZ	SA352	949006-2620	SCREW...................................1	
22 PAFZZ	PAFZZ	PAFZZ	SA352	949050-2590	NUT,HEXAGON..........................1	
23 PAFZZ	PAFZZ	PAFZZ	SA352	90258-10001	WASHER,LOCK...........................1	
24 XBFZZ	XB	XBFZZ	SA352	028065-0250	BUSHING,DRAIN.........................1	
25 PAFZZ	PAFZZ	PAFZZ	SA352	153400-0810	SWITCH ASSEMBLY,MAG.................1	
26 PAFZZ	PAFZZ	PAFZZ	SA352	949170-4070	SPRING,COMPRESSION..................1	
27 PAFZZ	PAFZZ	PAFZZ	SA352	949120-0240	BALL,STEEL..............................1	
28 XBFZZ	XB	XBFZZ	SA352	028300-3691	CLUTCH ASSEMBLY......................1	
29 XBFZZ	XB	XBFZZ	SA352	128401-8300	HOUSING.................................1	
30 PAFZZ	PAFZZ	PAFZZ	SA352	949006-4720	SCREW...................................3	
31 PAFZZ	PAFZZ	PAFZZ	75160	R56049	SPACER,STARTER.......................1	
32 PAFZZ	PAFZZ	PAFZZ	SA352	028099-3270	SERVICE KIT,PINION..................1	

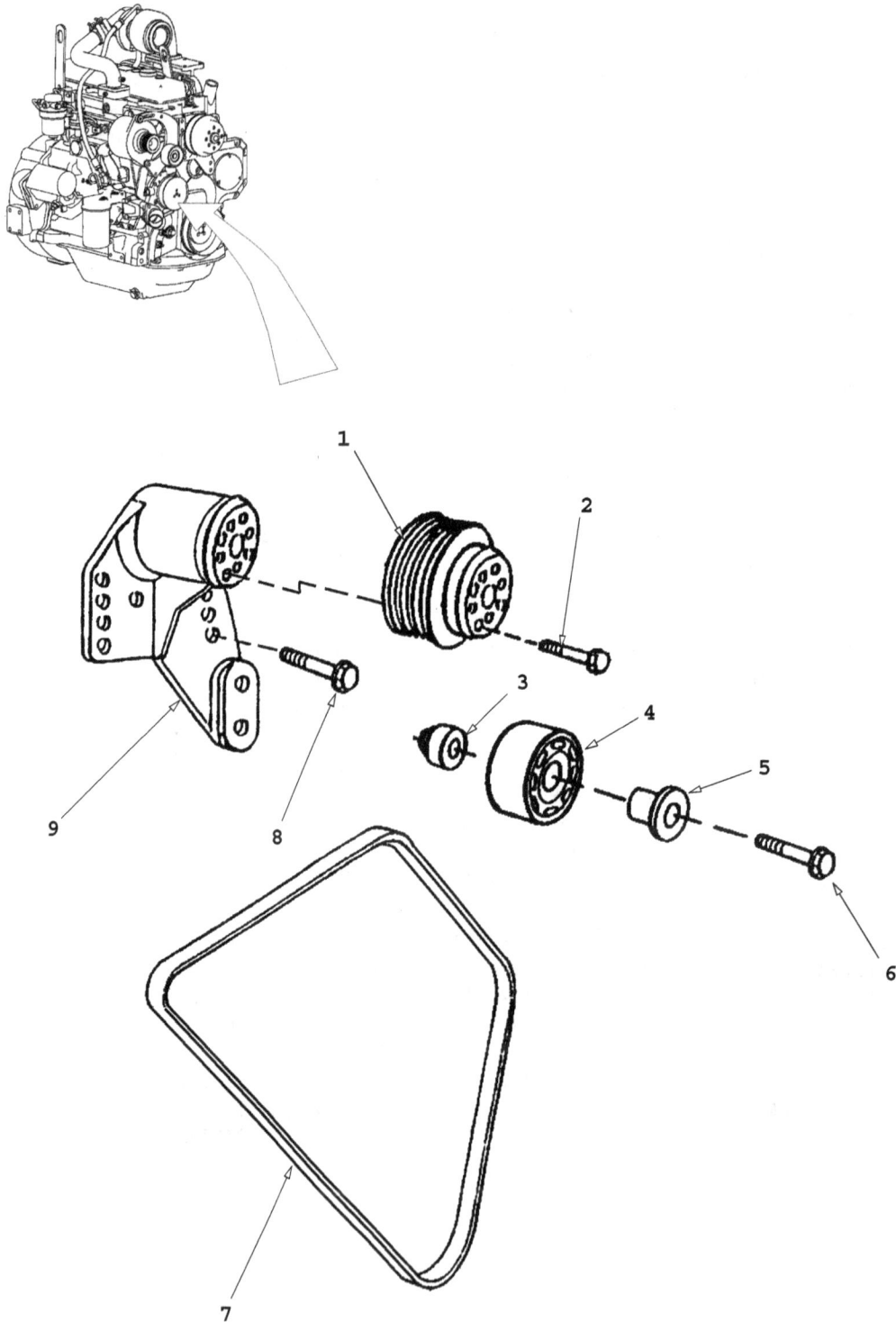

Figure 6. Fan Belt and Pulleys

(1) SMR CODE ITEM NO	ARMY	(2) AIR FORCE	USMC	(3) CAGEC	(4) PART NUMBER	(5) DESCRIPTION AND USABLE ON CODE (UOC)	(6) QTY

GROUP 03 INTAKE AND EXHAUST SYSTEM
FIGURE 6 FAN BELT AND PULLEYS

1	XBOZZ	XB	XBOZZ	66836	R128658	PULLEY................................	1
2	PAOZZ	PAOZZ	PAOZZ	66836	19M7784	BOLT,MACHINE........................	4
3	PAOZZ	PAOZZ	PAOZZ	66836	R500320	SPACER,STRAIGHT.....................	1
4	PAOZZ	PAOZZ	PAOZZ	66836	RE51281	PULLEY..............................	1
5	PAOZZ	PAOZZ	PAOZZ	66836	R107749	SLEEVE..............................	1
6	PAOZZ	PAOZZ	PAOZZ	66836	19M7807	BOLT,MACHINE........................	1
7	PAOZZ	PAOZZ	PAOZZ	66836	R501007	BELT,V..............................	1
8	PAOZZ	PAOZZ	PAOZZ	66836	19M7835	BOLT,MACHINE........................	4
9	PAOZZ	PAOZZ	PAOZZ	66836	RE500539	BEARING,HOUSING.....................	1

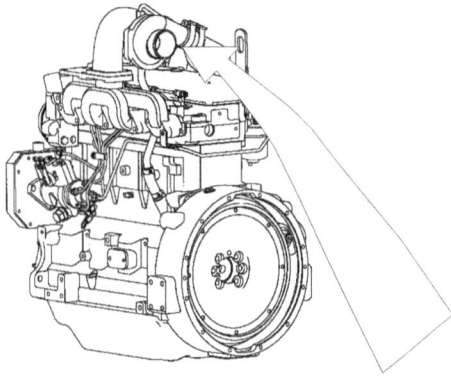

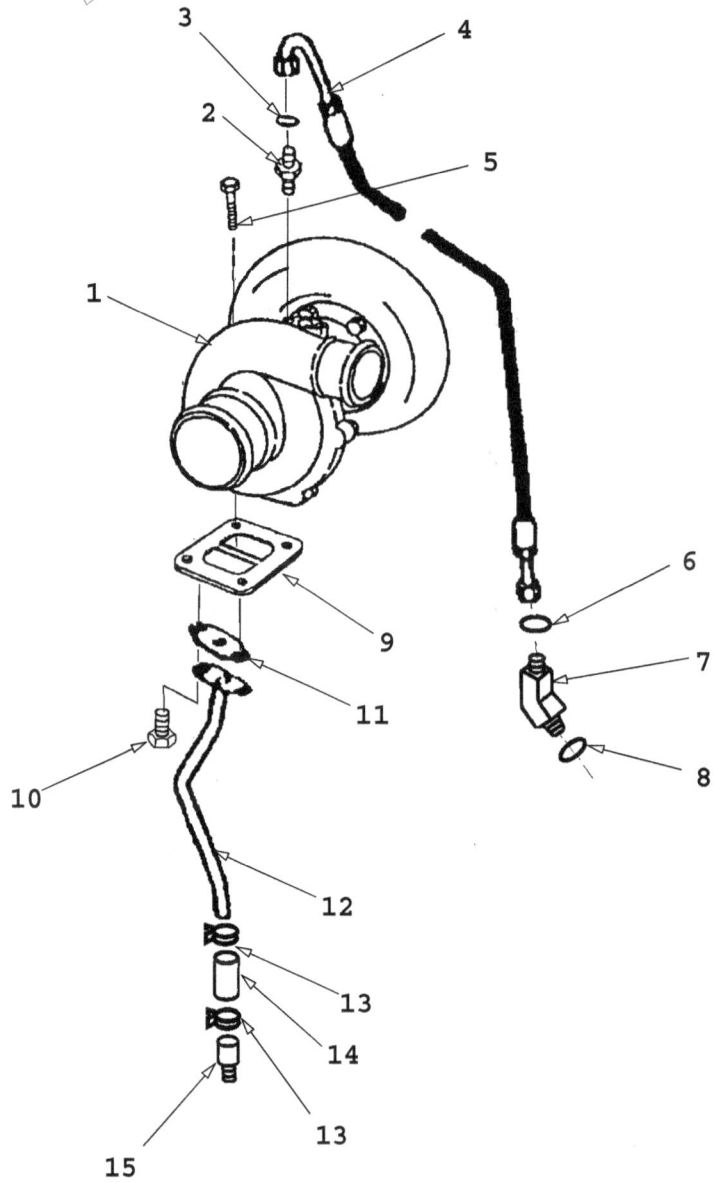

Figure 7. Turbocharger

(1) SMR CODE			(2)	(3)	(4)	(5)	(6)
ITEM NO	ARMY	AIR FORCE	USMC	CAGEC	PART NUMBER	DESCRIPTION AND USABLE ON CODE (UOC)	QTY

GROUP 0301 TURBOCHARGER
FIGURE 7 TURBOCHARGER

1	PAFZZ	PAFZZ	PAFZZ	66836	RE60076	TURBOCHARGER..........................1	
2	PAFZZ	PAFZZ	PAFZZ	66836	RE55871	ADAPTER,STRAIGHT......................1	
3	PAFZZ	PAFZZ	PAFZZ	66836	R63548	O-RING................................1	
4	PAFZZ	PAFZZ	PAFZZ	75160	RE59468	HOSE ASSEMBLY,NONME...................1	
5	PAFZZ	PAFZZ	PAFZZ	66836	19M7785	BOLT,MACHINE..........................4	
6	PAFZZ	PAFZZ	PAFZZ	75160	T77613	O-RING................................1	
7	PAFZZ	PAFZZ	PAFZZ	66836	38H5154	ELBOW.................................1	
8	PAFZZ	PAFZZ	PAFZZ	66836	51M7042	O-RING................................1	
9	PAFZZ	PAFZZ	PAFZZ	66836	R123572	GASKET................................1	
10	PAFZZ	PAFZZ	PAFZZ	75755	19M7866	BOLT,MACHINE..........................2	
11	PAFZZ	PAFZZ	PAFZZ	75160	R123570	GASKET................................1	
12	PAFZZ	PAFZZ	PAFZZ	66836	RE59547	HOSE..................................1	
13	PAFZZ	PAFZZ	PAFZZ	66836	RE65978	CLAMP,HOSE............................1	
14	PAFZZ	PAFZZ	PAFZZ	66836	R115138	HOSE..................................1	
15	PAFZZ	PAFZZ	PAFZZ	66836	RE54885	ADAPTER,STRAIGHT......................1	

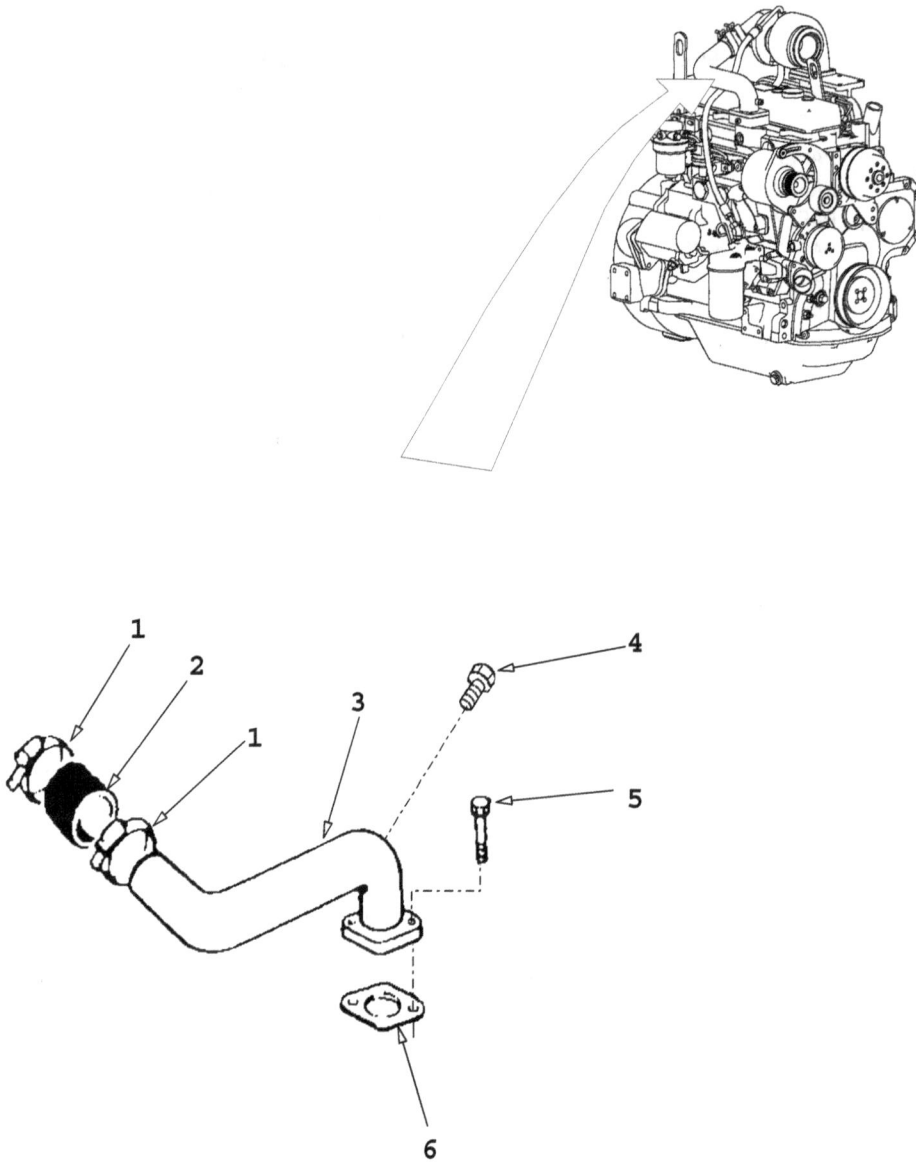

Figure 8. Intake Manifold

(1)			(3)	(4)	(5)	(6)
SMR CODE	(2)					
ITEM		AIR		PART	DESCRIPTION AND	
NO	ARMY	FORCE	USMC	CAGEC	NUMBER	USABLE ON CODE (UOC)	QTY

GROUP 0302 INTAKE MANIFOLD
FIGURE 8 INTAKE MANIFOLD

1	PAFZZ	PAFZZ	PAFZZ	75160	R110531	CLAMP,HOSE............................2
2	PAFZZ	PAFZZ	PAFZZ	75755	R56551	HOSE,NONMETALLIC.....................1
3	XBFZZ	XB	XBFZZ	66836	RE59341	TUBE,AIR INLET.......................1
4	XBFZZ	XB	XBFZZ	66836	RE60071	PLUG.................................1
5	PAFZZ	PAFZZ	PAFZZ	75755	19M7785	BOLT,MACHINE.........................2
6	PAFZZ	PAFZZ	PAFZZ	75160	R81275	GASKET...............................1

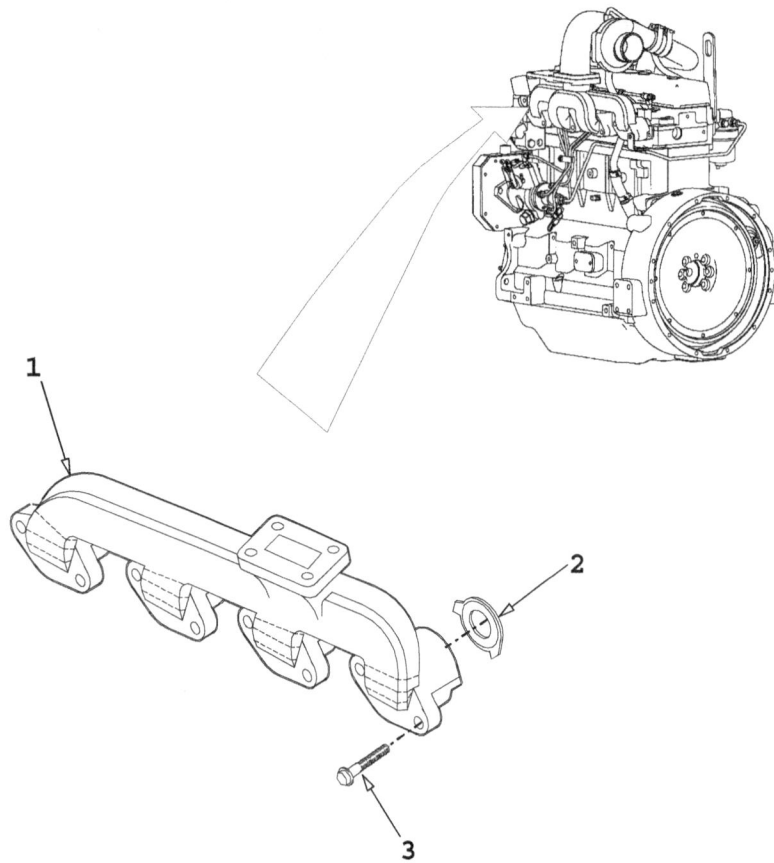

Figure 9.   Exhaust Manifold

(1)				(2)	(3)	(4)	(5)	(6)
SMR CODE								
ITEM		AIR				PART	DESCRIPTION AND	
NO	ARMY	FORCE	USMC	CAGEC		NUMBER	USABLE ON CODE (UOC)	QTY

GROUP 0303 EXHAUST MANIFOLD
FIGURE 9 EXHAUST MANIFOLD

1	XBFZZ	XB	XBFZZ	66836	R132373	MANIFOLD,EXHAUST.....................1
2	PAFZZ	PAFZZ	PAFZZ	75160	R90658	GASKET...............................4
3	PAFZZ	PAFZZ	PAFZZ	66836	19M7810	BOLT,MACHINE.........................8

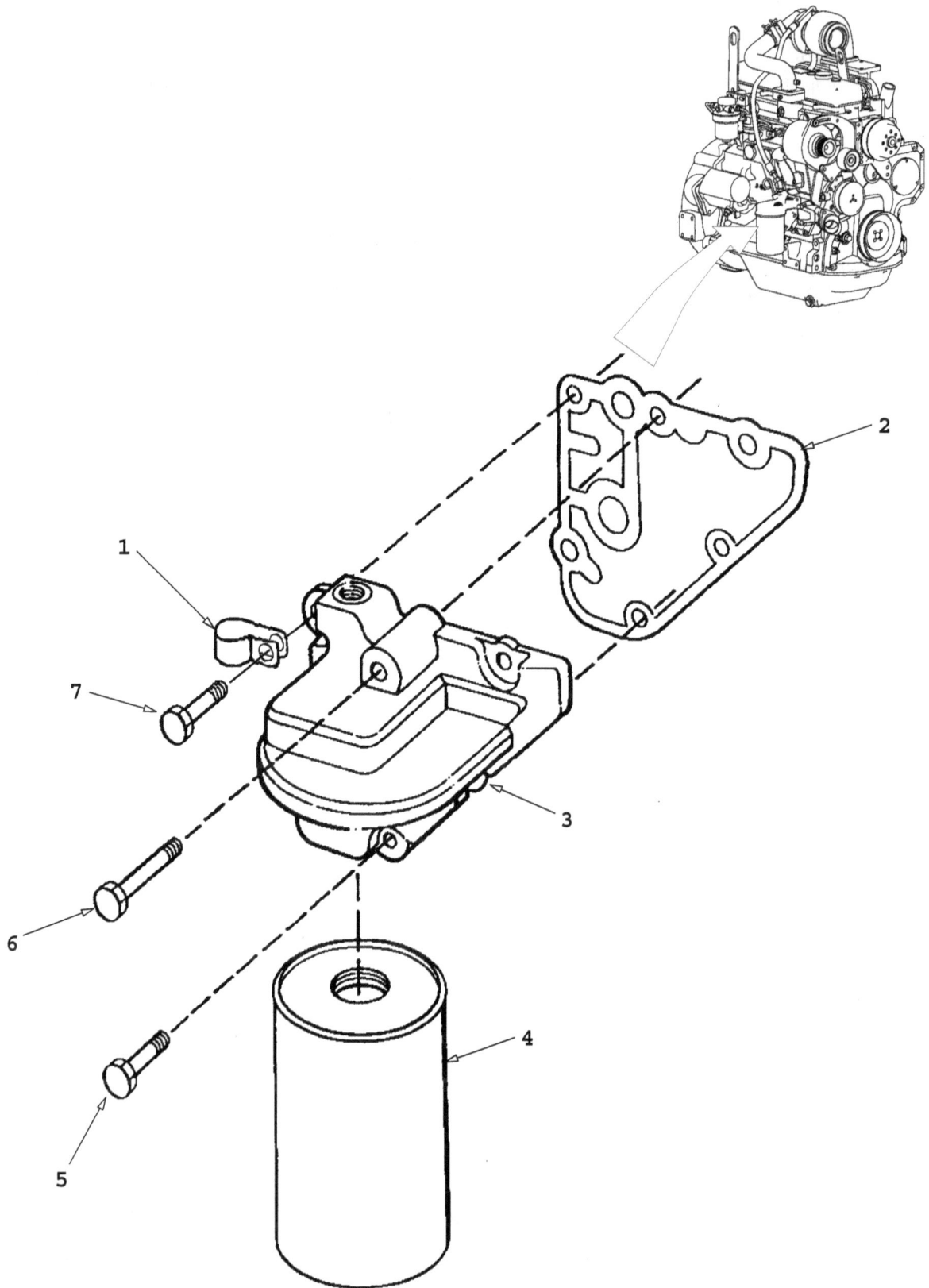

Figure 10. Oil Filter Assembly

(1) SMR CODE			(2)	(3)	(4)	(5)	(6)
ITEM NO	ARMY	AIR FORCE	USMC	CAGEC	PART NUMBER	DESCRIPTION AND USABLE ON CODE (UOC)	QTY

GROUP 0401 OIL FILTER ASSEMBLY
FIGURE 10 OIL FILTER ASSEMBLY

1	PAOZZ	PAOZZ	PAOZZ	66836	R123324	CLAMP,HOSE............................1	
2	PAOZZ	PAOZZ	PAOZZ	66836	R123525	GASKET................................1	
3	XBOZZ	XB	XBOZZ	66836	RE59907	HEADER,OIL FILTER.....................1	
4	PAOZZ	PAOZZ	PAOZZ	33968	36881696	FILTER ELEMENT,FLUI,OIL...............1	
5	PAOZZ	PAOZZ	PAOZZ	66836	19M7802	SCREW,FLANGED.........................3	
6	PAOZZ	PAOZZ	PAOZZ	75755	19M7866	BOLT,MACHINE..........................1	
7	PAOZZ	PAOZZ	PAOZZ	66836	19M7970	BOLT,MACHINE..........................1	

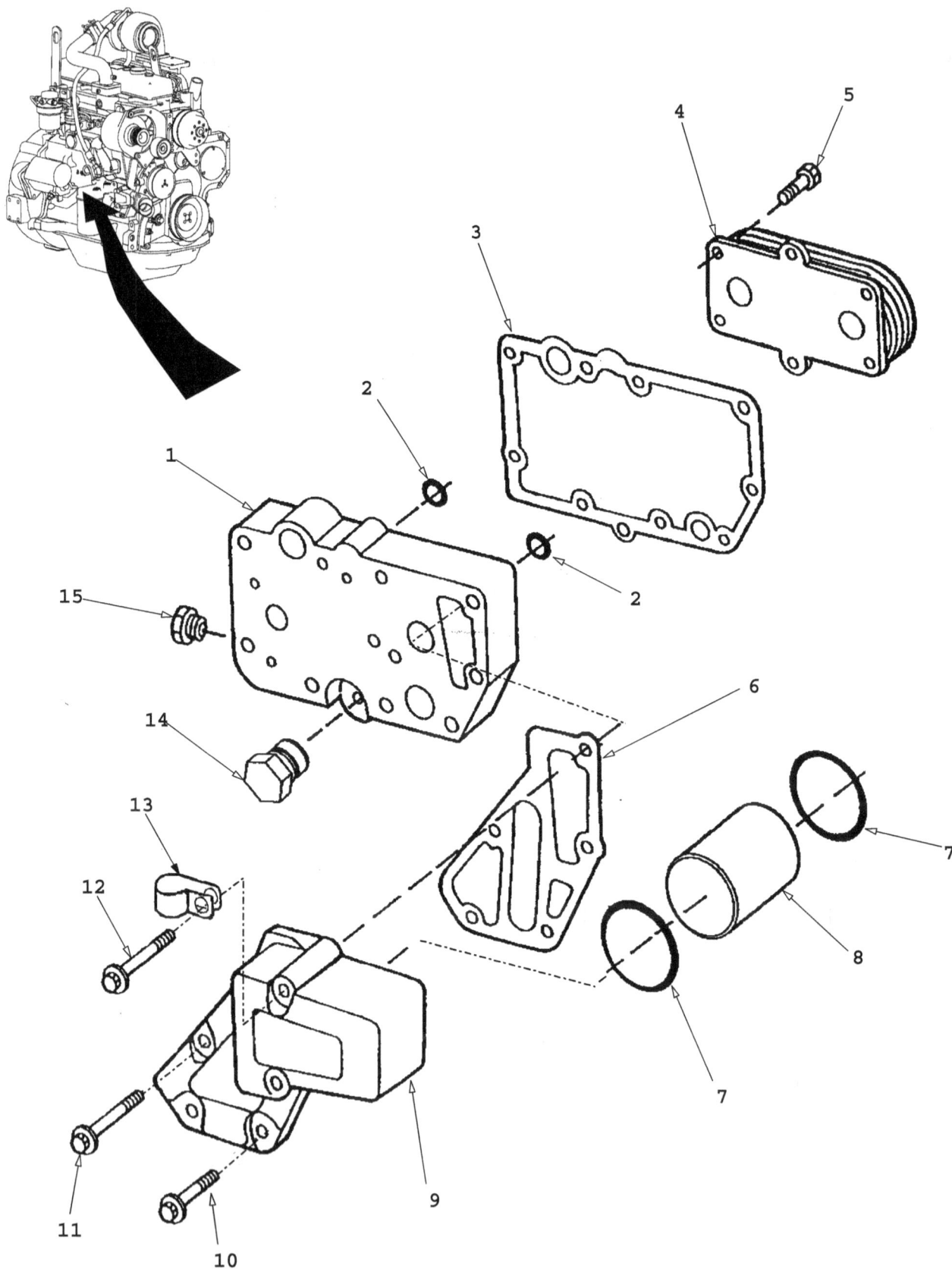

Figure 11. Oil Cooler

(1) SMR CODE			(3)	(4)	(5)	(6)
ITEM NO	(2)		CAGEC	PART NUMBER	DESCRIPTION AND USABLE ON CODE (UOC)	QTY
	ARMY	AIR FORCE	USMC			

GROUP 0402 OIL COOLER
FIGURE 11 OIL COOLER

ITEM NO	ARMY	AIR FORCE	USMC	CAGEC	PART NUMBER	DESCRIPTION	QTY
1	XBOZZ	XB	XBOZZ	66836	R123471	HOUSING,OIL COOLER...................1
2	PAOZZ	PAOZZ	PAOZZ	66836	T122075	O-RING.............................2
3	PAOZZ	PAOZZ	PAOZZ	66836	R123501	GASKET,HOUSING......................1
4	PAOZZ	PAOZZ	PAOZZ	66836	RE59812	COOLER,LUBRICATING..................1
5	PAOZZ	PAOZZ	PAOZZ	66836	19M8553	BOLT,SOCKET.........................6
6	PAOZZ	PAOZZ	PAOZZ	66836	R123502	GASKET.............................1
7	PAOZZ	PAOZZ	PAOZZ	66836	R61105	O-RING.............................2
8	XBOZZ	XB	XBOZZ	66836	R115252	TUBE,CONNECTOR......................1
9	XBOZZ	XB	XBOZZ	66836	R135177	ADAPTER............................1
10	PAOZZ	PAOZZ	PAOZZ	75755	19M7866	BOLT,MACHINE........................1
11	PAOZZ	PAOZZ	PAOZZ	66836	19M7802	SCREW,FLANGED.......................2
12	PAOZZ	PAOZZ	PAOZZ	66836	19M7970	BOLT,MACHINE........................2
13	PAOZZ	PAOZZ	PAOZZ	66836	R123324	CLAMP,HOSE..........................1
14	XBOZZ	XB	XBOZZ	66836	RE46685	PLUG...............................1
15	XBOZZ	XB	XBOZZ	66836	RE46686	PLUG...............................1

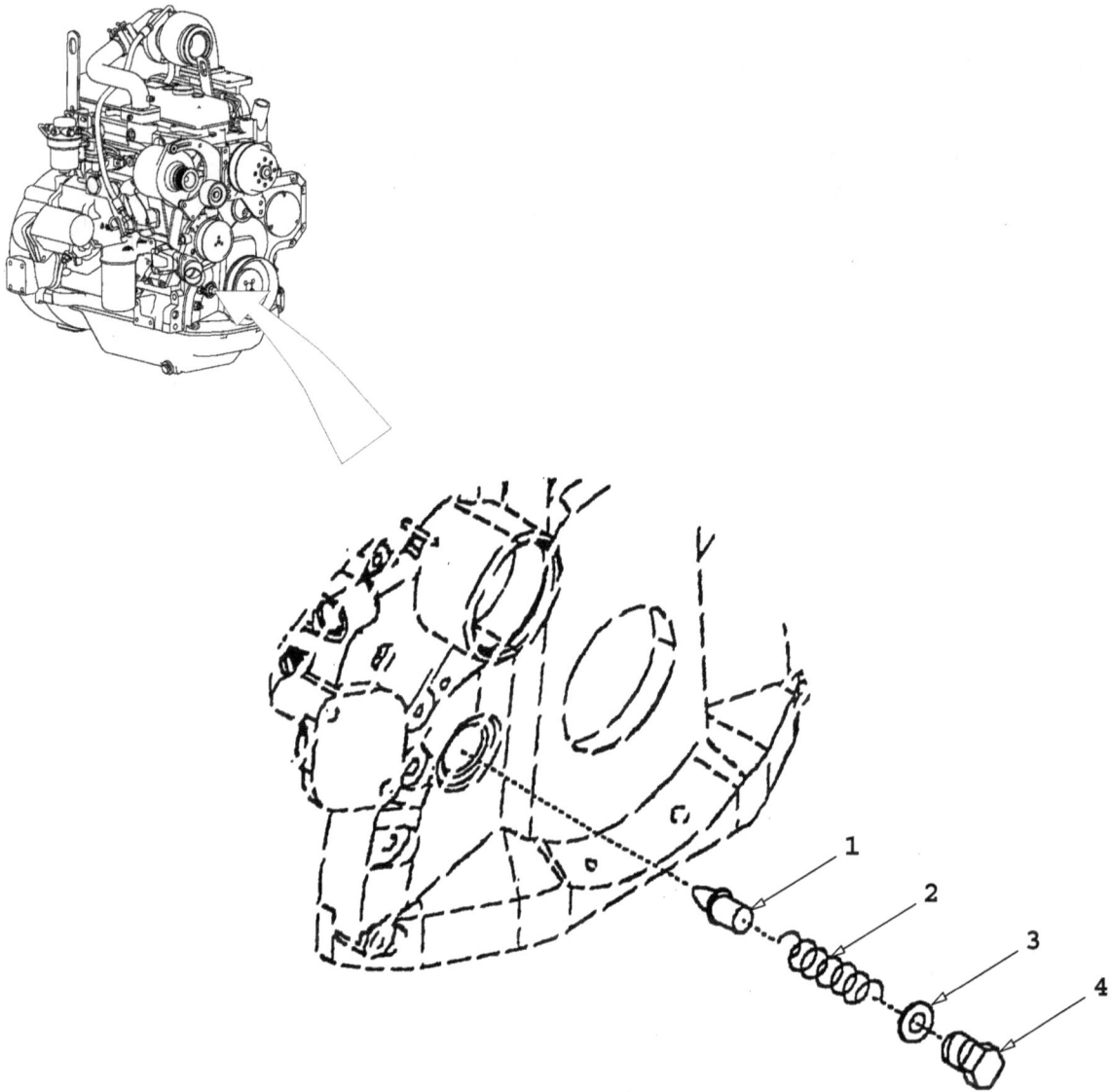

Figure 12. Oil Pressure Regulating Valve Assembly

(1) SMR CODE			(3)	(4)	(5)	(6)
ITEM NO	(2)		CAGEC	PART NUMBER	DESCRIPTION AND USABLE ON CODE (UOC)	QTY
	ARMY	AIR FORCE	USMC			

GROUP 0403 OIL PRESSURE REGULATING VALVE ASSEMBLY
FIGURE 12 OIL PRESSURE REGULATING VALVE ASSEMBLY

1	PAFZZ	PAFZZ	PAFZZ	75755	R83169	VALVE,FUEL SYSTEM.....................1
2	PAFZZ	PAFZZ	PAFZZ	66836	R131425	SPRING,HELICAL COMP..................1
3	PAFZZ	PAFZZ	PAFZZ	66836	A4827	WASHER,FLAT..........................1
4	XBFZZ	XB	XBFZZ	75755	R91692	PLUG,PIPE............................1

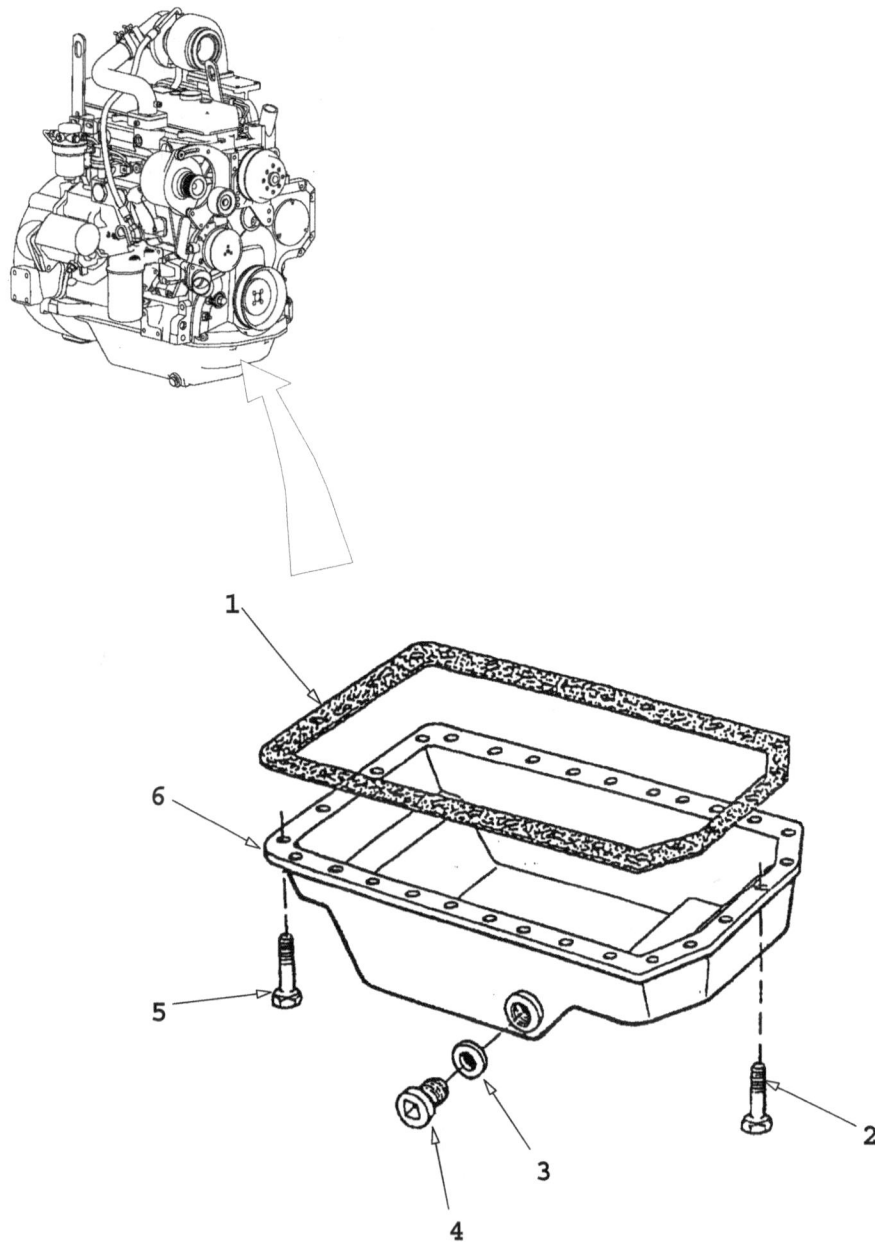

Figure 13. Oil Pan

(1) SMR CODE			(2)	(3)	(4)	(5)	(6)
ITEM NO	ARMY	AIR FORCE	USMC	CAGEC	PART NUMBER	DESCRIPTION AND USABLE ON CODE (UOC)	QTY

GROUP 0404 OIL PAN
FIGURE 13 OIL PAN

ITEM NO	ARMY	AIR FORCE	USMC	CAGEC	PART NUMBER	DESCRIPTION AND USABLE ON CODE (UOC)	QTY
1	PAFZZ	PAFZZ	PAFZZ	75160	R123352	GASKET................................1	
2	PAFZZ	PAFZZ	PAFZZ	66836	19M8062	SCREW,SERRATED,FLAN.................6	
3	PAFZZ	PAFZZ	PAFZZ	75160	51M7043	O-RING..............................1	
4	XBFZZ	XB	XBFZZ	66836	R107770	PLUG................................1	
5	PAFZZ	PAFZZ	PAFZZ	75755	19M8163	BOLT,MACHINE.......................22	
6	XBFZZ	XB	XBFZZ	66836	RE501022	PAN,OIL.............................1	

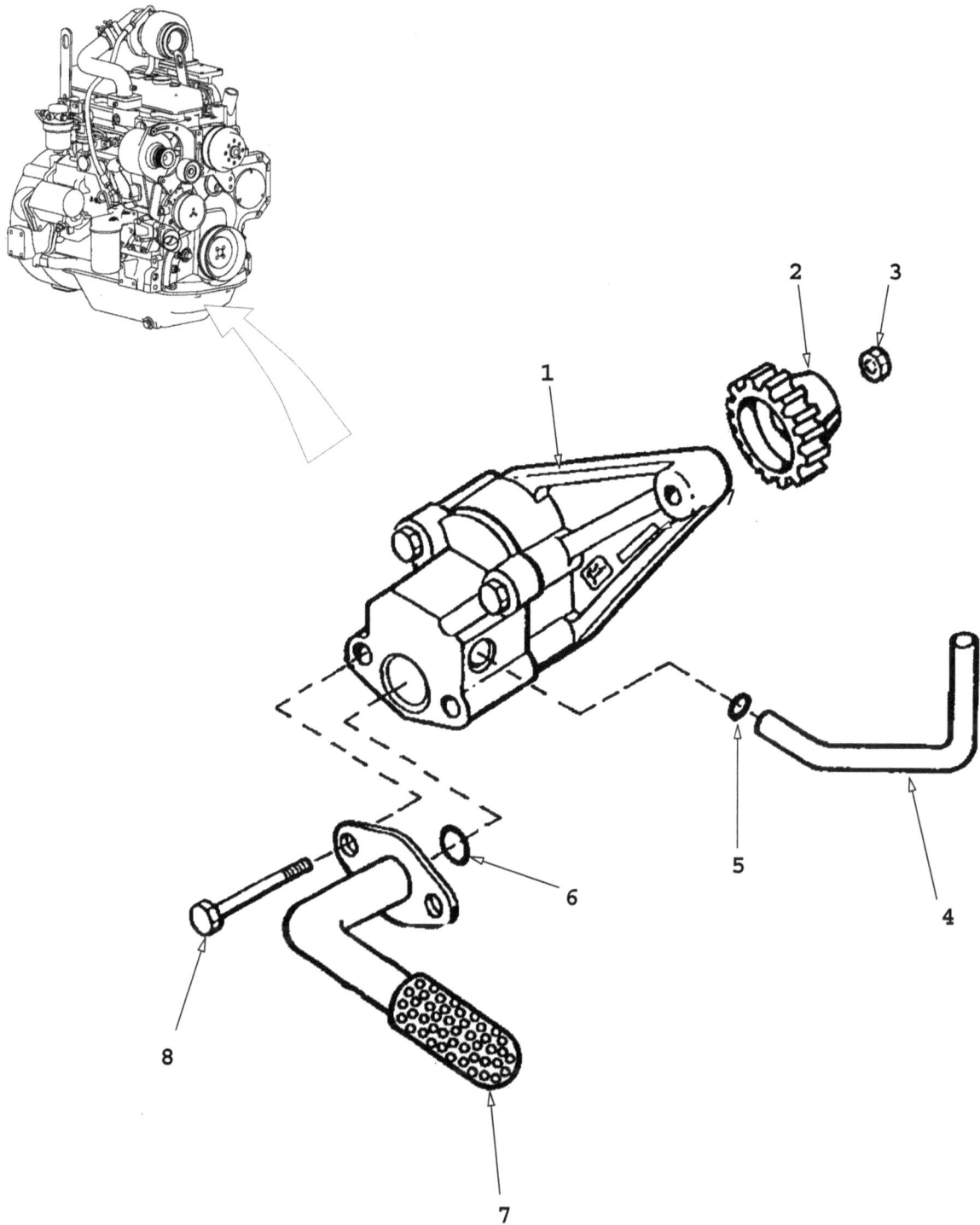

Figure 14. Oil Pump Assembly

(1) SMR CODE			(2)	(3)	(4)	(5)	(6)
ITEM NO	ARMY	AIR FORCE	USMC	CAGEC	PART NUMBER	DESCRIPTION AND USABLE ON CODE (UOC)	QTY

GROUP 0405 OIL PUMP ASSEMBLY
FIGURE 14 OIL PUMP ASSEMBLY

1	PAFZZ	PAFZZ	PAFZZ	66836	RE502269	PUMP,OIL...........................1	
2	PAFZZ	PAFZZ	PAFZZ	66836	R120638	GEAR,SPUR...........................1	
3	PAFZZ	PAFZZ	PAFZZ	66836	14M7066	NUT,HEX.............................1	
4	XBFZZ	XB	XBFZZ	66836	R121376	TUBE,OUTLET.........................1	
5	PAFZZ	PAFZZ	PAFZZ	66836	R97185	O-RING..............................1	
6	PAFZZ	PAFZZ	PAFZZ	75160	R61871	O-RING..............................1	
7	XBFZZ	XB	XBFZZ	66836	RE57619	INTAKE,OIL..........................1	
8	PAFZZ	PAFZZ	PAFZZ	66836	19M7970	BOLT,MACHINE........................2	

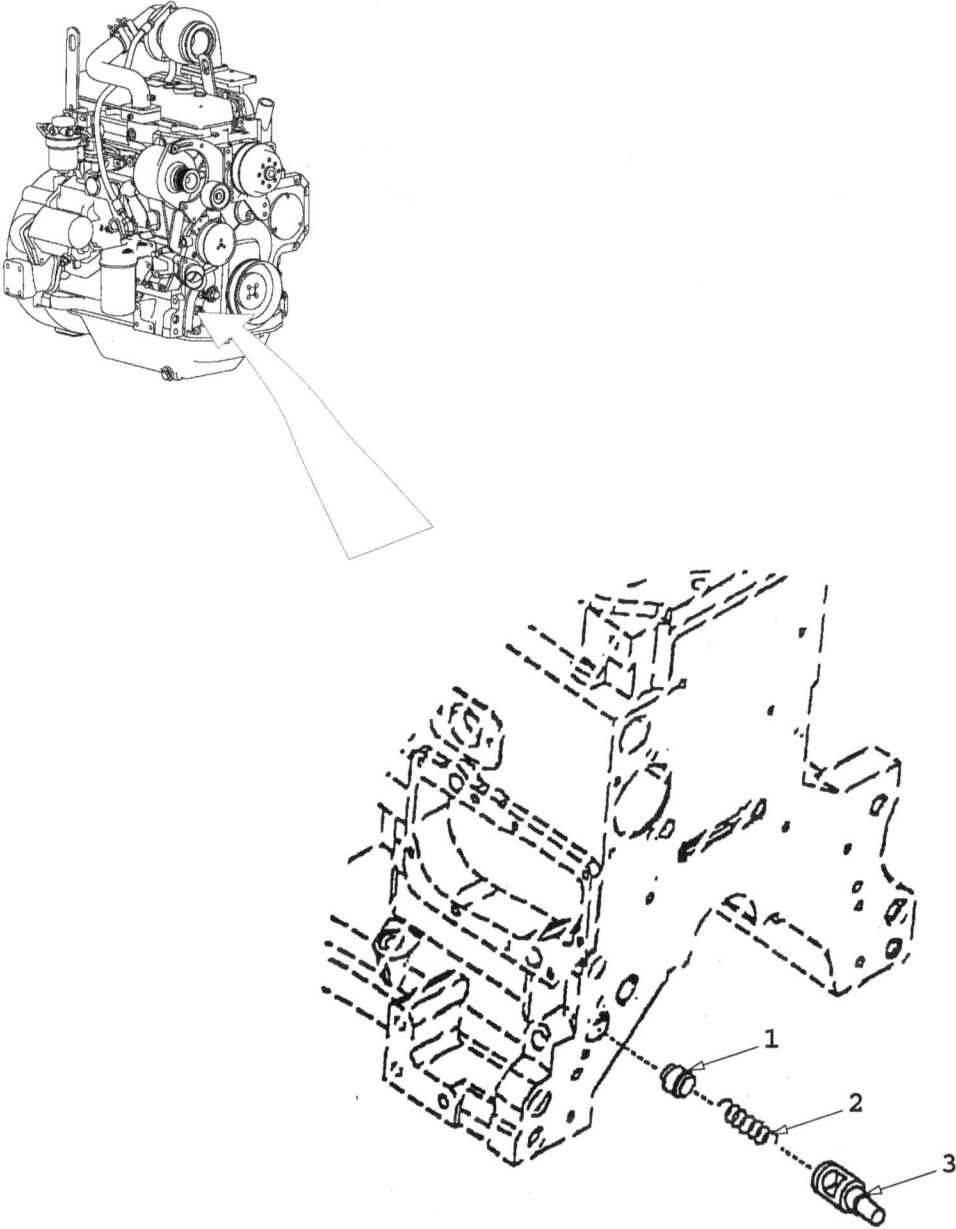

Figure 15. Oil Bypass Valve Assembly

(1)				(3)	(4)	(5)	(6)
SMR CODE							
ITEM		AIR			PART	DESCRIPTION AND	
NO	ARMY	FORCE	USMC	CAGEC	NUMBER	USABLE ON CODE (UOC)	QTY

GROUP 0406 OIL BYPASS VALVE ASSEMBLY
FIGURE 15 OIL BYPASS VALVE ASSEMBLY

1	PAFZZ	PAFZZ	PAFZZ	75160	R26493	BUSHING,SLEEVE........................1
2	PAFZZ	PAFZZ	PAFZZ	75160	R111137	SPRING,HELICAL,COMP...................1
3	PAFZZ	PAFZZ	PAFZZ	66836	R121043	VALVE,REGULATING,BYPASS...............1

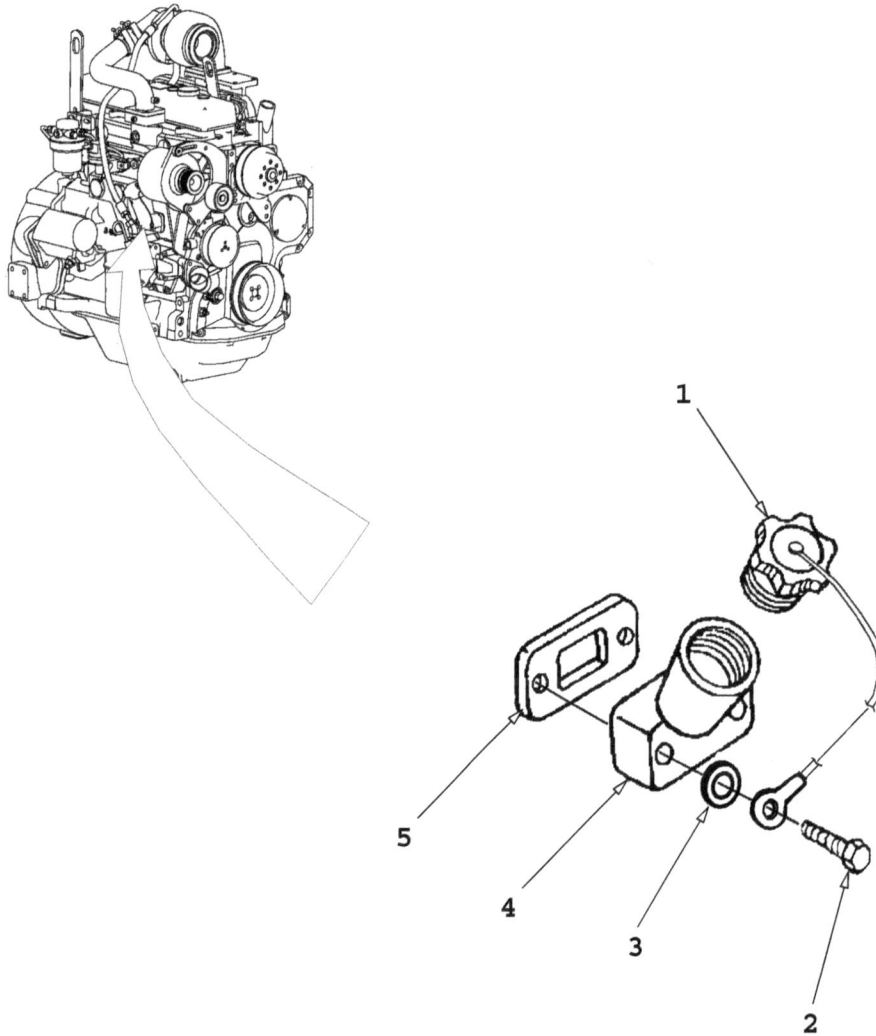

Figure 16. Oil Fill Tube

(1)			(2)	(3)	(4)	(5)	(6)
SMR CODE							
ITEM		AIR			PART	DESCRIPTION AND	
NO	ARMY	FORCE	USMC	CAGEC	NUMBER	USABLE ON CODE (UOC)	QTY

GROUP 0407 OIL FILL TUBE
FIGURE 16 OIL FILL TUBE

1	PAOZZ	PAOZZ	PAOZZ	66836	RE501472	CAP,FILLER OPENING...................1	
2	PAOZZ	PAOZZ	PAOZZ	66836	19M7979	BOLT,MACHINE.........................2	
3	PAOZZ	PAOZZ	PAOZZ	66836	RE71526	WASHER...............................2	
4	XBOZZ	XB	XBOZZ	66836	R136483	FILLER NECK..........................1	
5	PAOZZ	PAOZZ	PAOZZ	66836	R136495	GASKET...............................1	

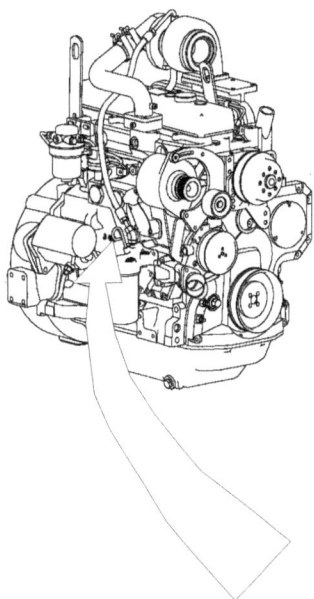

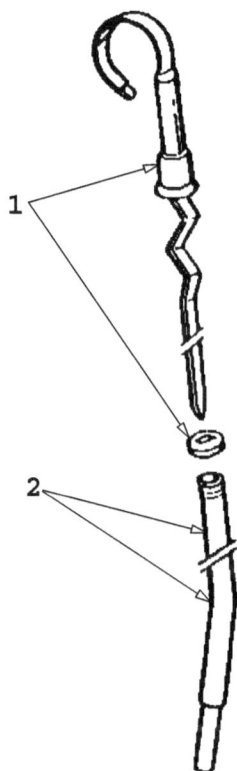

Figure 17. Oil Dipstick Tube

(1) SMR CODE ITEM NO	ARMY	(2) AIR FORCE	USMC	(3) CAGEC	(4) PART NUMBER	(5) DESCRIPTION AND USABLE ON CODE (UOC)	(6) QTY

GROUP 0408 OIL DIPSTICK TUBE
FIGURE 17 OIL DIPSTICK TUBE

1	PAOZZ	PAOZZ	PAOZZ	66836	RE69243	GAGE ROD,LIQUID LEVEL,OIL............	1
2	XBOZZ	XB	XBOZZ	66836	R500709	TUBE,GAGE ROD........................	1

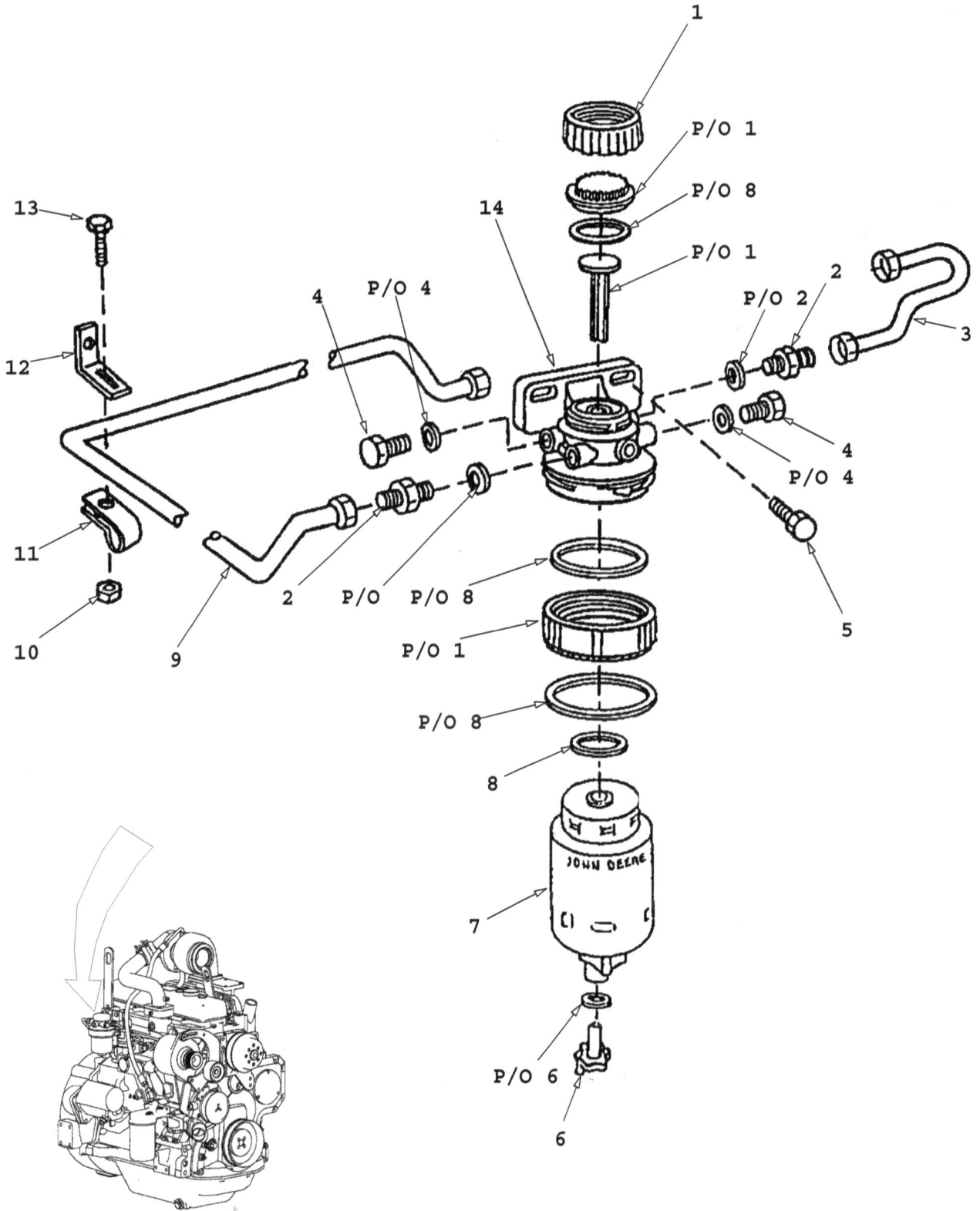

Figure 18. Fuel Filter

(1)		(2)		(3)	(4)	(5)	(6)
SMR CODE							
ITEM		AIR			PART	DESCRIPTION AND	
NO	ARMY	FORCE	USMC	CAGEC	NUMBER	USABLE ON CODE (UOC)	QTY

GROUP 0501 FUEL FILTER
FIGURE 18 FUEL FILTER

1	PAOZZ	PAOZZ	PAOZZ	75755	RE51649	PARTS KIT,SEAL REPL..................1	
2	PAOZZ	PAOZZ	PAOZZ	66836	RE60029	ADAPTER,STRAIGHT....................2	
3	XBOZZ	XB	XBOZZ	66836	RE60037	LINE,FUEL...........................1	
4	XBOZZ	XB	XBOZZ	66836	RE46686	PLUG................................1	
5	PAOZZ	PAOZZ	PAOZZ	75755	19M7785	BOLT,MACHINE........................2	
6	PAOZZ	PAOZZ	PAOZZ	66836	RE60854	KIT,VALVE...........................1	
7	PAOZZ	PAOZZ	PAOZZ	75160	RE60021	FILTER ELEMENT,FLUI.................1	
8	PAOZZ	PAOZZ	PAOZZ	75755	RE50752	PARTS KIT,SEAL REPL.................1	
9	XBOZZ	XB	XBOZZ	66836	RE60025	TUBE,METALLIC.......................1	
10	PAOZZ	PAOZZ	PAOZZ	75755	14M7298	NUT,HEX,FLANGED.....................1	
11	PAOZZ	PAOZZ	PAOZZ	66836	T14050	CLAMP...............................1	
12	XBOZZ	XB	XBOZZ	66836	R116116	BRACKET.............................1	
13	PAOZZ	PAOZZ	PAOZZ	66836	19M7865	BOLT,MACHINE........................1	
14	XBOZZ	XB	XBOZZ	66836	RE60022	HEADER,FILTER.......................1	

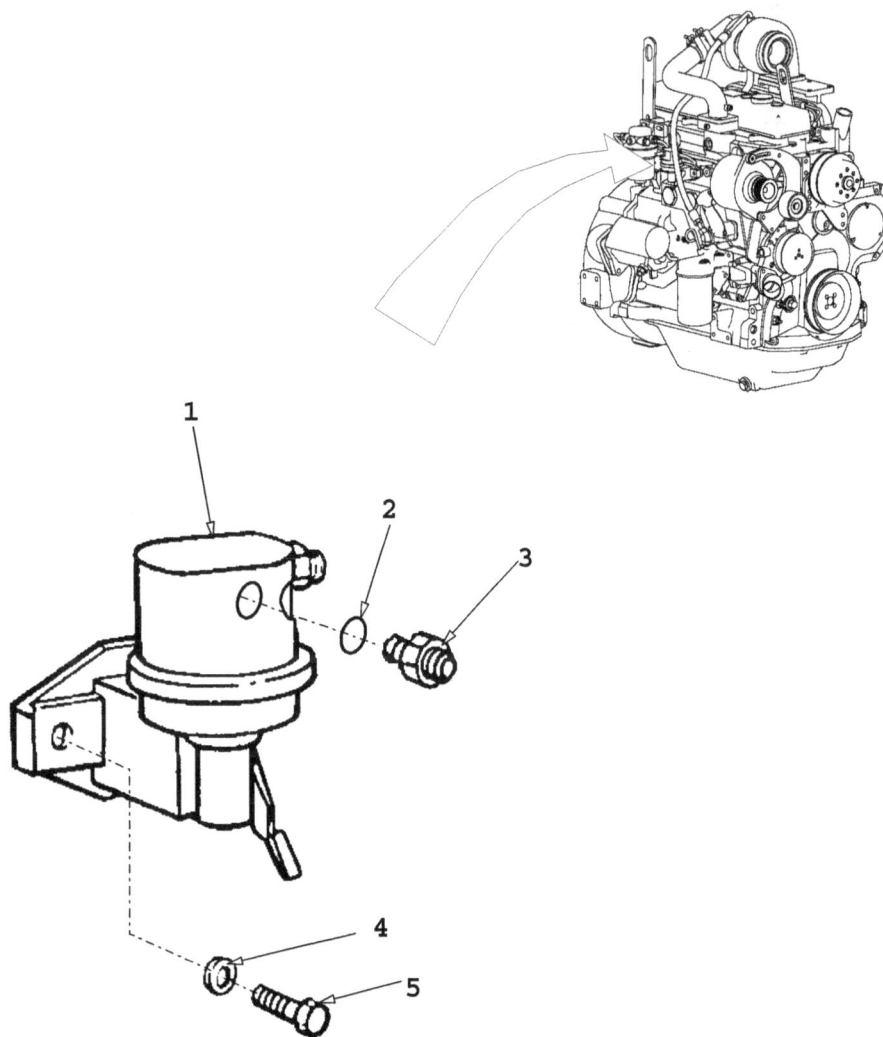

Figure 19. Fuel Transfer Supply Pump

(1) SMR CODE				(3)	(4)	(5)	(6)
ITEM NO	ARMY	AIR FORCE	(2) USMC	CAGEC	PART NUMBER	DESCRIPTION AND USABLE ON CODE (UOC)	QTY

GROUP 0502 FUEL TRANSFER SUPPLY PUMP
FIGURE 19 FUEL TRANSFER SUPPLY PUMP

1	PAOZZ	PAOZZ	PAOZZ	66836	RE68345	PUMP,FUEL,TRANSFER....................	1
2	PAOZZ	PAOZZ	PAOZZ	66836	R500271	O-RING...............................	1
3	PAOZZ	PAOZZ	PAOZZ	66836	R500404	ADAPTER,HOSE.........................	1
4	PAOZZ	PAOZZ	PAOZZ	66836	24M7055	WASHER...............................	2
5	PAOZZ	PAOZZ	PAOZZ	66836	RE66298	SCREW,CAP............................	2

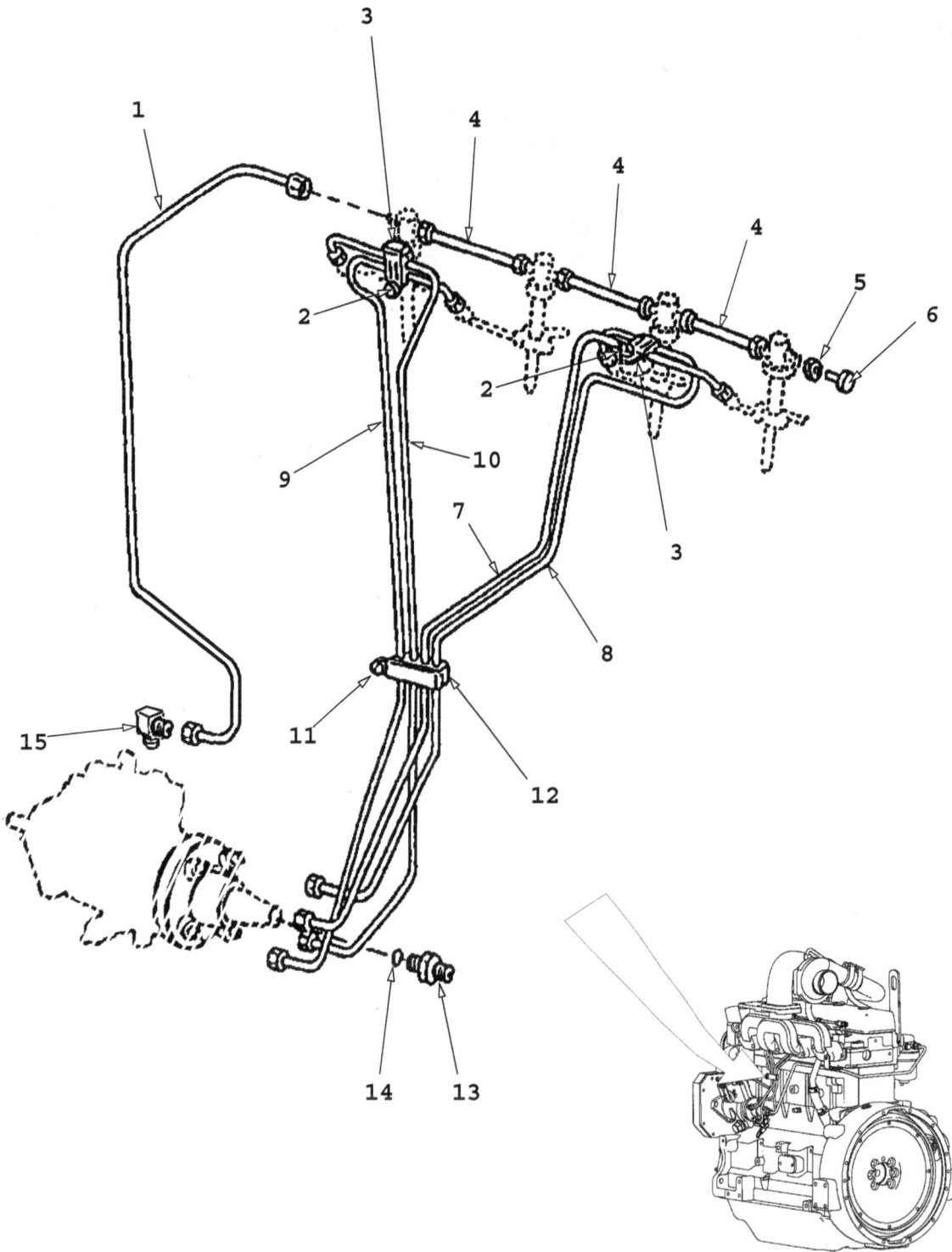

Figure 20. Fuel Lines

(1)				(4)	(5)	(6)
SMR CODE	(2)		(3)			
ITEM	AIR			PART	DESCRIPTION AND	
NO	ARMY	FORCE	USMC	CAGEC	NUMBER	DESCRIPTION AND USABLE ON CODE (UOC)	QTY

GROUP 0503 FUEL LINES
FIGURE 20 FUEL LINES

1	PAOZZ	PAOZZ	PAOZZ	66836	RE70848	HOSE,METALLIC,LEAK-OFF...............1
2	PAOZZ	PAOZZ	PAOZZ	75160	RE19797	CLAMP,LOOP..........................4
3	PAOZZ	PAOZZ	PAOZZ	66836	19M6614	SCREW,CAP,HEXAGON H..................1
4	PAOZZ	PAOZZ	PAOZZ	66836	RE68748	HOSE,METALLIC,LEAK-OFF..............4
5	PAOZZ	PAOZZ	PAOZZ	66836	R123593	NUT,PLAIN,HEXAGON...................2
6	XBFZZ	XB	XBFZZ	75755	AR85519	PLUG,PIPE...........................2
7	PAOZZ	PAOZZ	PAOZZ	66836	RE59450	HOSE,METALLIC ,FUEL INJECTION NO. 4................1
8	PAOZZ	PAOZZ	PAOZZ	66836	RE59449	HOSE,METALLIC ,FUEL INJECTION NO. 3................1
9	PAOZZ	PAOZZ	PAOZZ	66836	RE59448	HOSE,METALLIC ,FUEL INJECTION NO. 2................1
10	PAOZZ	PAOZZ	PAOZZ	66836	RE59447	HOSE,METALLIC ,FUEL INJECTION NO. 1................1
11	PAFZZ	PAFZZ	PAFZZ	75160	RE19799	CLAMP,LOOP..........................1
12	PAOZZ	PAOZZ	PAOZZ	66836	19M6614	SCREW,CAP,HEXAGON H..................1
13	PAOZZ	PAOZZ	PAOZZ	66836	RE60029	ADAPTER,STRAIGHT....................1
14	PAOZZ	PAOZZ	PAOZZ	66836	51M7040	O-RING..............................1
15	PAOZZ	PAOZZ	PAOZZ	75755	R67364	ELBOW,PIPE TO TUBE..................1

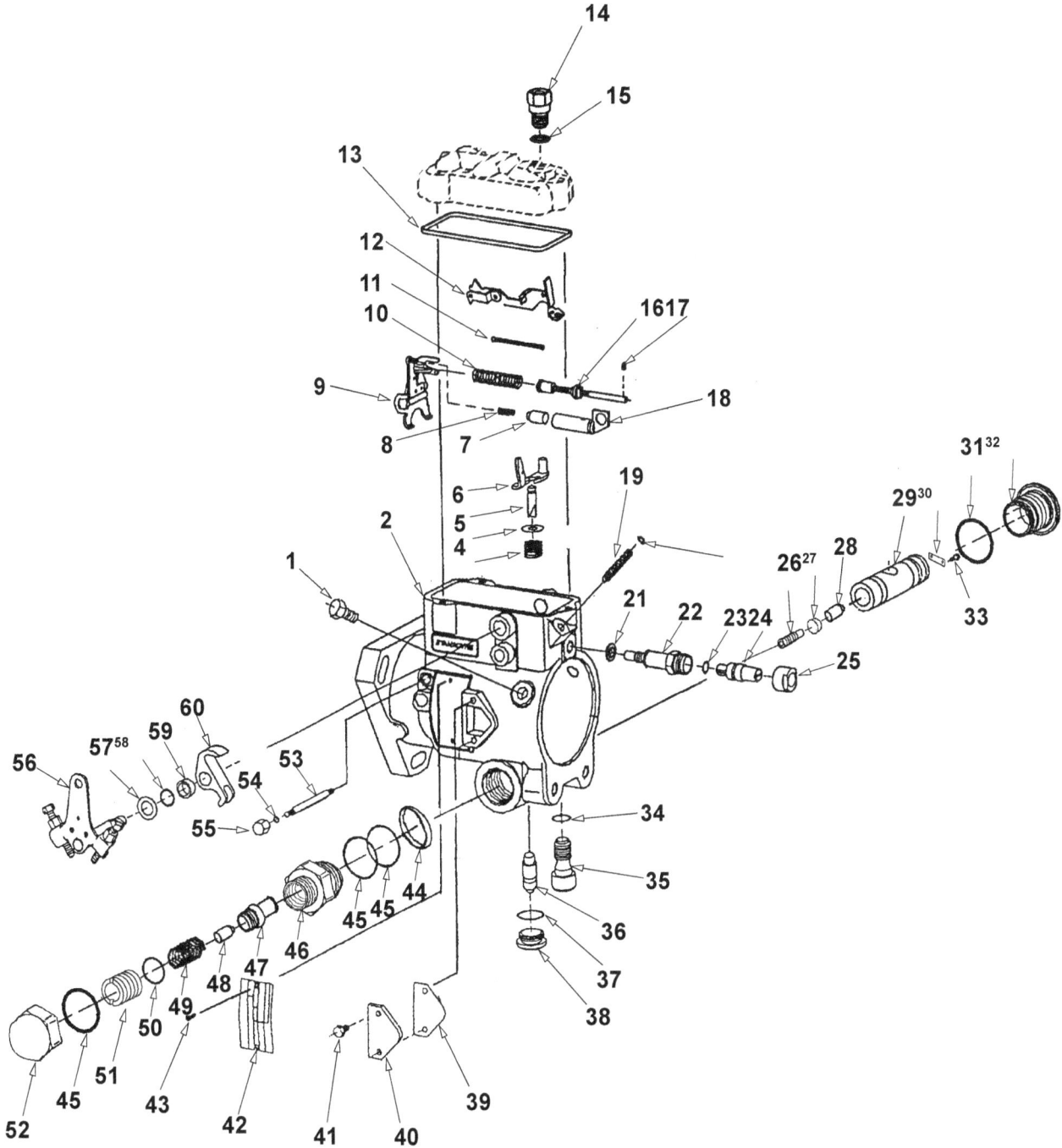

Figure 21. Fuel Injection Pump (Sheet 1 of 2)

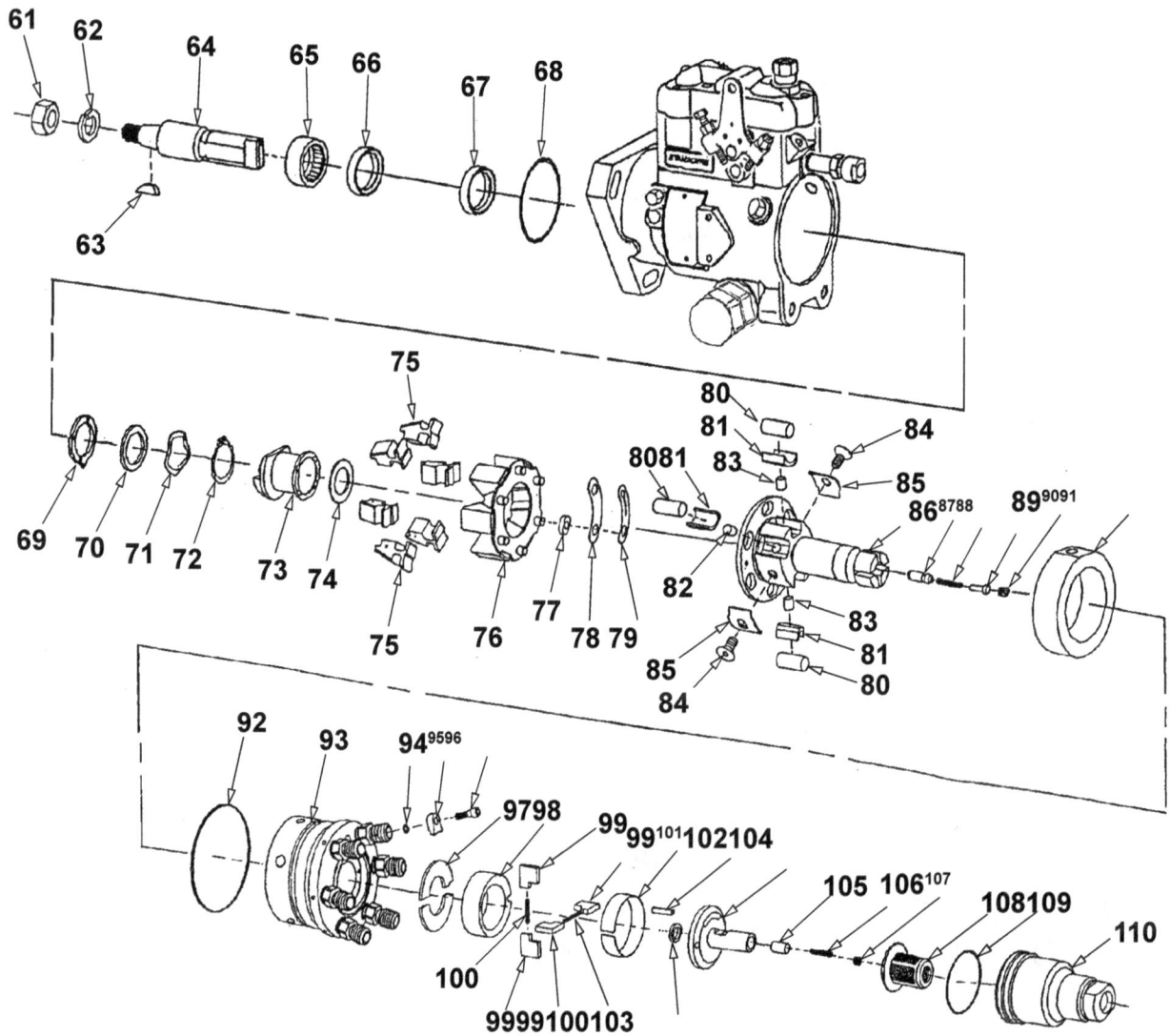

Figure 21.   Fuel Injection Pump (Sheet 2 of 2)

(1) SMR CODE ITEM NO	ARMY	(2) AIR FORCE	USMC	(3) CAGEC	(4) PART NUMBER	(5) DESCRIPTION AND USABLE ON CODE (UOC)	(6) QTY

GROUP 0504 FUEL INJECTION PUMP
FIGURE 21 FUEL INJECTION PUMP

Item	Army	AF	USMC	CAGEC	Part No	Description	Qty
1	XBHZZ	XB	XBHZZ	84760	11331	SCREW,HEXAGON HEAD...................1	
2	XBHZZ	XB	XBHZZ	84760	32478	HOUSING ASSEMBLY......................1	
3	XBHZZ	XB	XBHZZ	84760	20359	SPRING.............................1	
4	XBHZZ	XB	XBHZZ	84760	26427	SHIM..............................1	
5	XBHZZ	XB	XBHZZ	84760	11563	VALVE,METERING.....................1	
6	XBHZZ	XB	XBHZZ	84760	22134	ARM ASSEMBLY.......................1	
7	XBHZZ	XB	XBHZZ	84760	16572	PISTON ASSEMBLY....................1	
8	XBHZZ	XB	XBHZZ	84760	20475	SPRING.............................1	
9	XBHZZ	XB	XBHZZ	84760	20956	ARM,GOVERNOR.......................1	
10	XBHZZ	XB	XBHZZ	84760	13560	SPRING.............................1	
11	XBHZZ	XB	XBHZZ	84760	11919	SPRING.............................1	
12	XBHZZ	XB	XBHZZ	84760	17604	HOOK,GOVERNOR LINKA................1	
13	XBHZZ	XB	XBHZZ	84760	27244	SEAL..............................1	
14	XBHZZ	XB	XBHZZ	84760	22109	REGULATOR,PRESSURE.................1	
15	XBHZZ	XB	XBHZZ	84760	27607	O-RING............................1	
16	XBHZZ	XB	XBHZZ	84760	20240	ROD ASSEMBLY,CONTRO................1	
17	XBHZZ	XB	XBHZZ	84760	13554	PIN,STRAIGHT,HEADLE................1	
18	XBHZZ	XB	XBHZZ	84760	16568	BARREL ASSEMBLY....................1	
19	XBHZZ	XB	XBHZZ	84760	22601	SCREW.............................1	
20	XBHZZ	XB	XBHZZ	84760	27606	WASHER,FLAT.......................1	
21	XBHZZ	XB	XBHZZ	84760	27606	WASHER,FLAT.......................1	
22	XBHZZ	XB	XBHZZ	84760	27599	GUIDE,ROD.........................1	
23	XBHZZ	XB	XBHZZ	84760	12966	O-RING............................1	
24	XBHZZ	XB	XBHZZ	84760	20243	CAP ASSEMBLY,ADJUST................1	
25	XBHZZ	XB	XBHZZ	84760	20355	CAP,LOCKING.......................1	
26	XBHZZ	XB	XBHZZ	84760	33840	SPRING............................1	
27	XBHZZ	XB	XBHZZ	84760	23197	SEAT,SPRING.......................1	
28	XBHZZ	XB	XBHZZ	84760	32466	VALVE,SERVO.......................1	
29	XBHZZ	XB	XBHZZ	84760	32519	PISTON............................1	
30	XBHZZ	XB	XBHZZ	84760	21146	VALVE.............................1	
31	XBHZZ	XB	XBHZZ	84760	32533	O-RING............................1	
32	XBHZZ	XB	XBHZZ	84760	22649	PLUG..............................1	
33	XBHZZ	XB	XBHZZ	84760	19834	SCREW.............................1	
34	XBHZZ	XB	XBHZZ	84760	27602	O-RING............................2	
35	XBHZZ	XB	XBHZZ	84760	22238	SCREW,ALLEN HD,SPEC................1	
36	XBHZZ	XB	XBHZZ	84760	21147	PIN...............................1	
37	XBHZZ	XB	XBHZZ	84760	27610	O-RING............................1	
38	XBHZZ	XB	XBHZZ	84760	23056	PLUG..............................1	
39	XBHZZ	XB	XBHZZ	84760	27603	GASKET............................1	
40	XBHZZ	XB	XBHZZ	84760	23107	COVER.............................1	

(1) SMR CODE			(2)	(3)	(4)	(5)	(6)
ITEM NO	ARMY	AIR FORCE	USMC	CAGEC	PART NUMBER	DESCRIPTION AND USABLE ON CODE (UOC)	QTY
41	XBHZZ	XB	XBHZZ	84760	21194	SCREW,HEXAGON HEAD....................2
42	XBHZZ	XB	XBHZZ	84760	10349	PLATE,IDENTIFICATIO...................1
43	XBHZZ	XB	XBHZZ	84760	24419	SCREW.................................1
44	XBHZZ	XB	XBHZZ	84760	32947	WASHER,TEFLON.........................1
45	XBHZZ	XB	XBHZZ	84760	32948	O-RING................................1
46	XBHZZ	XB	XBHZZ	84760	33334	PLUG ASSEMBLY.........................1
47	XBHZZ	XB	XBHZZ	84760	34033	PISTON................................1
48	XBHZZ	XB	XBHZZ	84760	34245	PLUG..................................1
49	XBHZZ	XB	XBHZZ	84760	34284	SPRING................................1
50	XBHZZ	XB	XBHZZ	84760	33613	O-RING................................1
51	XBHZZ	XB	XBHZZ	84760	33614	SCREW,HEXAGON HEAD....................1
52	XBHZZ	XB	XBHZZ	84760	33044	CAP,ADVANCE...........................1
53	XBHZZ	XB	XBHZZ	84760	12288	NUT,HEXAGON...........................2
54	XBHZZ	XB	XBHZZ	84760	31332	O-RING................................2
55	XBHZZ	XB	XBHZZ	84760	20224	SHAFT.................................1
56	XBHZZ	XB	XBHZZ	84760	16251	SHAFT ASSEMBLY........................1
57	XBHZZ	XB	XBHZZ	84760	14408	WASHER,FLAT...........................1
58	XBHZZ	XB	XBHZZ	84760	17438	O-RING................................1
59	XBHZZ	XB	XBHZZ	84760	15306	SPACER................................1
60	XBHZZ	XB	XBHZZ	84760	12221	LEVER.................................1
61	XBHZZ	XB	XBHZZ	84760	26346	NUT,HEXAGON...........................1
62	XBHZZ	XB	XBHZZ	84760	32787	WASHER,LOCK...........................1
63	XBHZZ	XB	XBHZZ	84760	30387	KEY,WOODRUFF..........................1
64	XBHZZ	XB	XBHZZ	84760	32880	SHAFT,DRIVE...........................1
65	XBHZZ	XB	XBHZZ	84760	26359	BEARING,NEEDLE........................1
66	XBHZZ	XB	XBHZZ	84760	26355	SEAL,OIL..............................1
67	XBHZZ	XB	XBHZZ	84760	30804	SEAL,OIL..............................1
68	XBHZZ	XB	XBHZZ	84760	26470	SEAL..................................1
69	XBHZZ	XB	XBHZZ	84760	28749	BEARING,THRUST........................1
70	XBHZZ	XB	XBHZZ	84760	26358	WASHER,FLAT...........................1
71	XBHZZ	XB	XBHZZ	84760	28381	WASHER,LOCK...........................1
72	XBHZZ	XB	XBHZZ	84760	26361	RING,RETAINING........................1
73	XBHZZ	XB	XBHZZ	84760	21709	SLEEVE................................1
74	XBHZZ	XB	XBHZZ	84760	23856	WASHER,FLAT...........................1
75	XBHZZ	XB	XBHZZ	84760	28974	WEIGHT,GOVERNOR.......................6
76	XBHZZ	XB	XBHZZ	84760	23860	RETAINER,WEIGHT.......................1
77	XBHZZ	XB	XBHZZ	84760	19873	CUSHION...............................6
78	XBHZZ	XB	XBHZZ	84760	19894	RETAINER..............................3
79	XBHZZ	XB	XBHZZ	84760	29136	CLIP,RETAINING........................3
80	XBHZZ	XB	XBHZZ	84760	11141	ROLLER,CAM............................4
81	XBHZZ	XB	XBHZZ	84760	20120	SHOE,ROLLER...........................4
82	XBHZZ	XB	XBHZZ	84760	18700	PLUNGER...............................1
83	XBHZZ	XB	XBHZZ	84760	11056	PLUNGER...............................1
84	XBHZZ	XB	XBHZZ	84760	15619	SCREW,ALLEN HEAD......................2
85	XBHZZ	XB	XBHZZ	84760	29363	SPRING,LEAF...........................2

(1)		(2)		(3)	(4)	(5)	(6)
SMR CODE ITEM NO	ARMY	AIR FORCE	USMC	CAGEC	PART NUMBER	DESCRIPTION AND USABLE ON CODE (UOC)	QTY
86	XBHZZ	XB	XBHZZ	84760	29366	ROTOR ASSEMBLY........................1	
87	XBHZZ	XB	XBHZZ	84760	21213	VALVE,DELIVERY........................1	
88	XBHZZ	XB	XBHZZ	84760	20129	SPRING................................1	
89	XBHZZ	XB	XBHZZ	84760	20128	STOP,VALVE............................1	
90	XBHZZ	XB	XBHZZ	84760	20127	SCREW.................................1	
91	XBHZZ	XB	XBHZZ	84760	32163	CAM RING..............................1	
92	XBHZZ	XB	XBHZZ	84760	27245	O-RING................................1	
93	XBHZZ	XB	XBHZZ	84760	34322	HEAD ASSEMBLY,HYDRA...................1	
94	XBHZZ	XB	XBHZZ	84760	27601	O-RING................................1	
95	XBHZZ	XB	XBHZZ	84760	29384	PLATE,METAL...........................1	
96	XBHZZ	XB	XBHZZ	84760	29710	SCREW,ALLEN HEAD......................1	
97	XBHZZ	XB	XBHZZ	84760	32859	RETAINER..............................1	
98	XBHZZ	XB	XBHZZ	84760	22988	LINER,PUMP............................1	
99	XBHZZ	XB	XBHZZ	84760	20803	PLATE,METAL...........................4	
100	XBHZZ	XB	XBHZZ	84760	15699	SPRING................................2	
101	XBHZZ	XB	XBHZZ	84760	19906	RING..................................1	
102	XBHZZ	XB	XBHZZ	84760	19837	PIN,STRAIGHT,HEADLE...................1	
103	XBHZZ	XB	XBHZZ	84760	19844	SEAL..................................1	
104	XBHZZ	XB	XBHZZ	84760	32548	REGULATOR ASSEMBLY....................1	
105	XBHZZ	XB	XBHZZ	84760	19895	PISTON,REGULATOR......................1	
106	XBHZZ	XB	XBHZZ	84760	23915	SPRING................................1	
107	XBHZZ	XB	XBHZZ	84760	32642	PLUG ASSEMBLY.........................1	
108	XBHZZ	XB	XBHZZ	84760	28986	FILTER................................1	
109	XBHZZ	XB	XBHZZ	84760	27608	O-RING................................1	
110	XBHZZ	XB	XBHZZ	84760	32636	CAP...................................1	

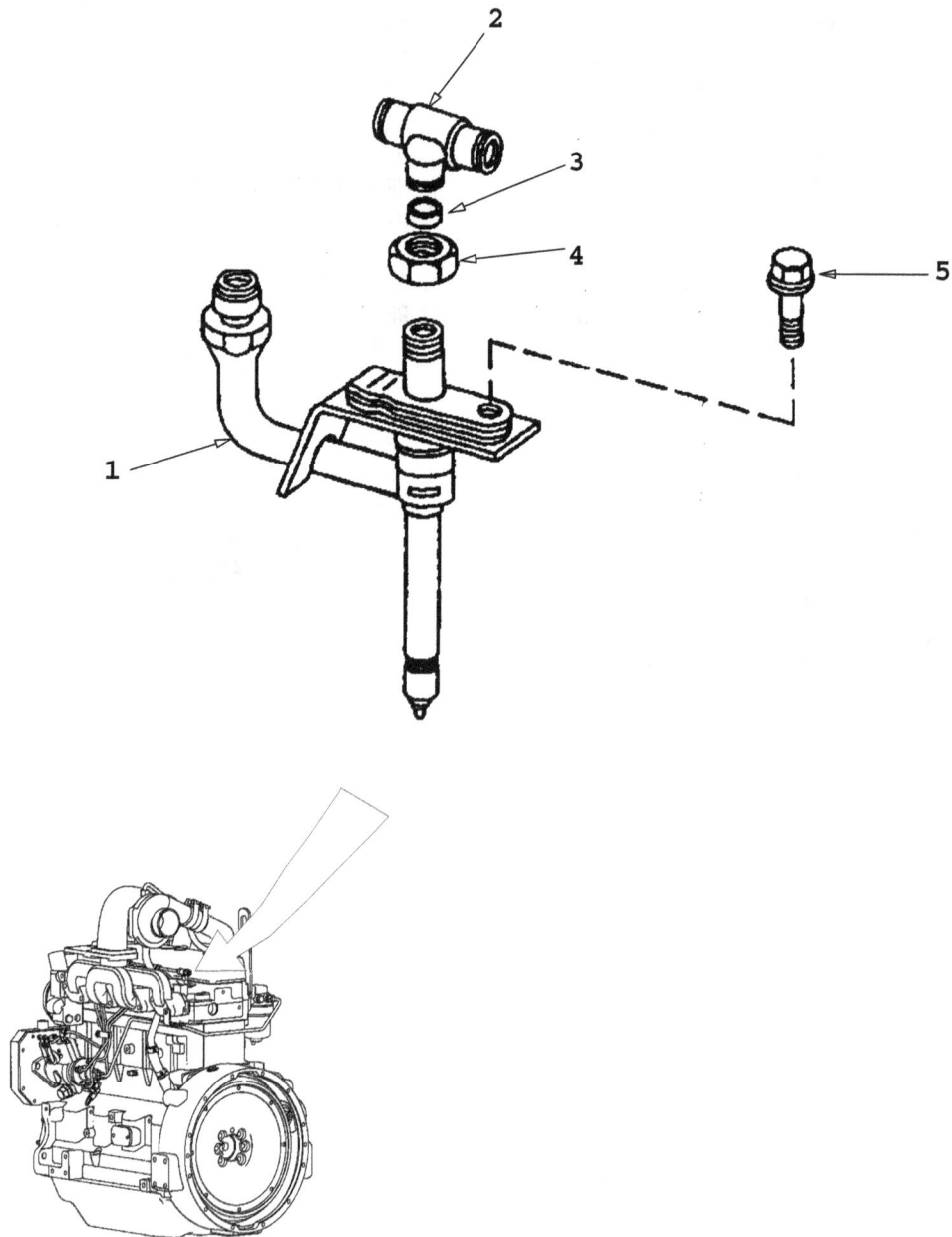

Figure 22. Fuel Injection Nozzle Assembly

(1) SMR CODE			(2)		(3)	(4)	(5)	(6)
ITEM NO	ARMY		AIR FORCE	USMC	CAGEC	PART NUMBER	DESCRIPTION AND USABLE ON CODE (UOC)	QTY

GROUP 0505 FUEL INJECTION NOZZLE ASSEMBLY
FIGURE 22 FUEL INJECTION NOZZLE ASSEMBLY

1	PAFZZ		PAFBZ	PAFZZ	66836	RE48786	NOZZLE,FUEL INJECTI................4	
2	XBFZZ		XB	XBFZZ	66836	R71963	FITTING,TEE.........................1	
3	PAFZZ		PAFZZ	PAFZZ	66836	R79605	WASHER.............................1	
4	PAFZZ		PAFZZ	PAFZZ	66836	R79604	NUT................................1	
5	PAFZZ		PAFZZ	PAFZZ	66836	R116366	SCREW,SPECIAL......................1	

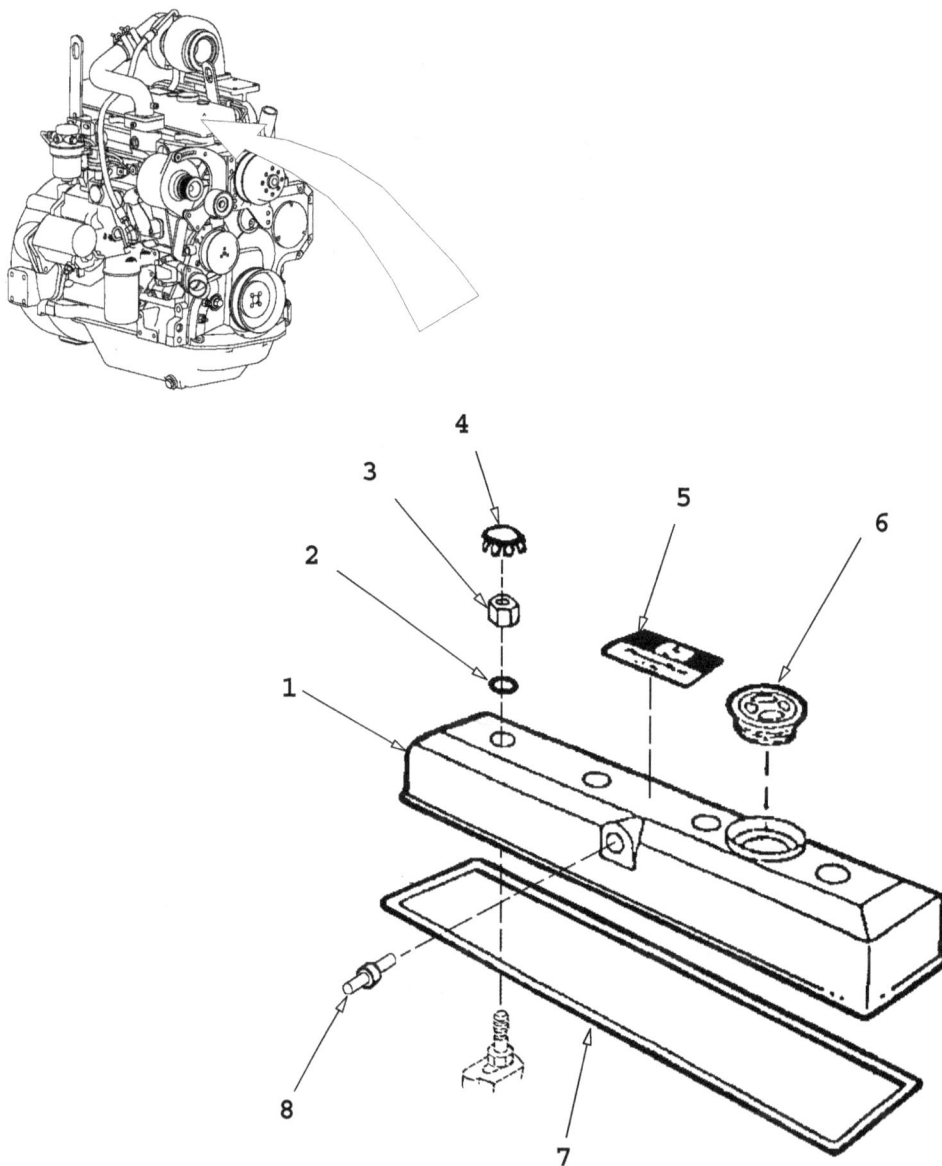

Figure 23. Rocker Arm Cover

(1)				(4)	(5)	(6)
SMR CODE						
ITEM NO	ARMY	AIR FORCE	USMC	CAGEC	PART NUMBER	DESCRIPTION AND USABLE ON CODE (UOC)	QTY

GROUP 0601 ROCKER ARM COVER
FIGURE 23 ROCKER ARM COVER

1	XBFZZ	XB	XBFZZ	66836	RE70401	COVER,ENGINE POPPET..................1
2	PAFZZ	PAFZZ	PAFZZ	75160	R123575	O-RING..............................4
3	PAFZZ	PAFZZ	PAFZZ	66836	R123574	NUT,HEX,FLANGED.....................4
4	PAFZZ	PAFZZ	PAFZZ	66836	H23125	PLUG,BUTTON.........................4
5	XBFZZ	XB	XBFZZ	66836	R116324	PLATE,IDENTIFICATIO.................1
6	XBFZZ	XB	XBFZZ	66836	RE59801	PLUG,MACHINE THREAD.................1
7	PAFZZ	PAFZZ	PAFZZ	75160	R123543	GASKET..............................1
8	PAFZZ	PAFZZ	PAFZZ	66836	R502572	FITTING.............................1

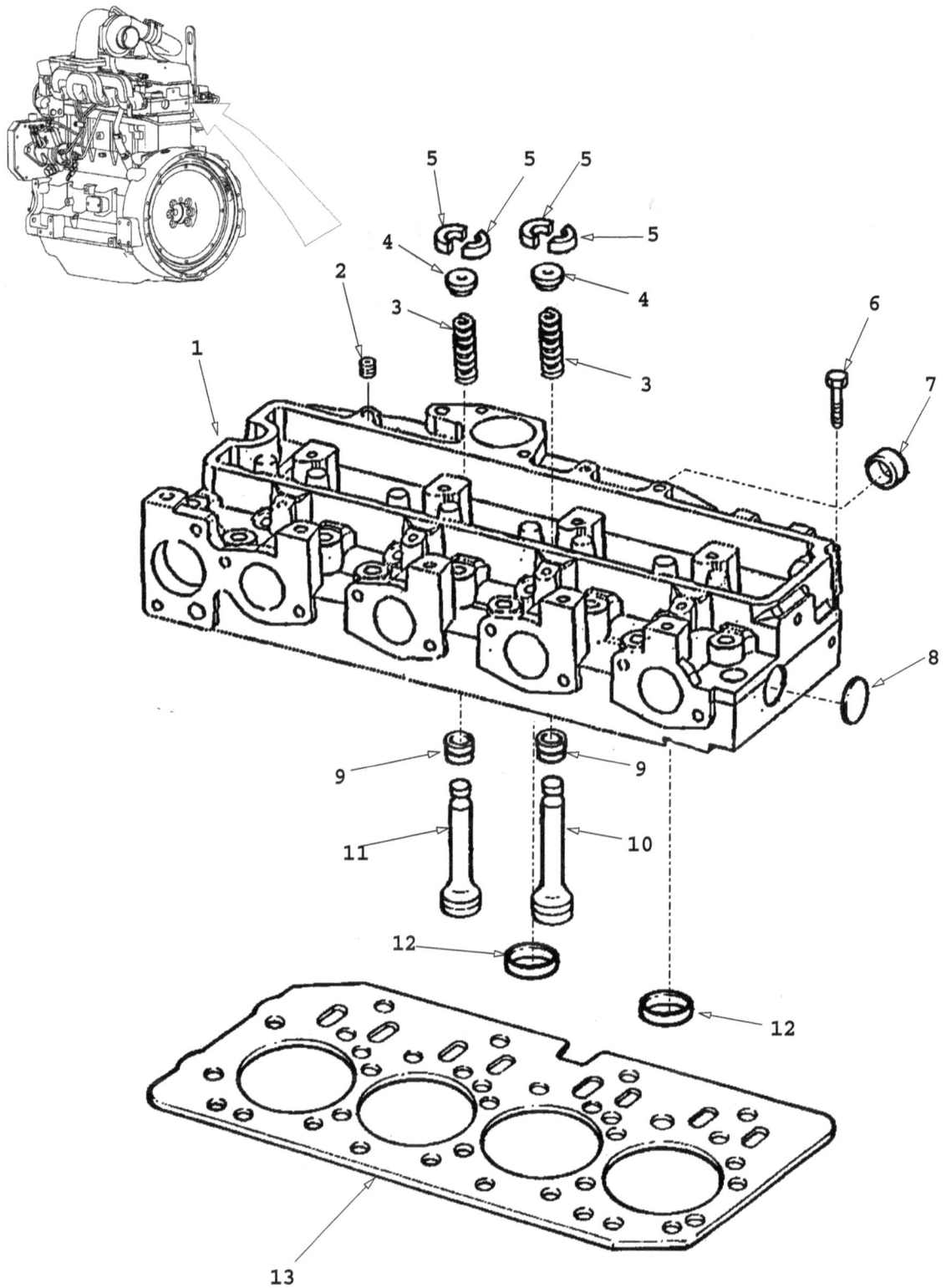

Figure 24. Cylinder Head, Intake, and Exhaust Valves

(1) SMR CODE			(2)	(3)	(4)	(5)	(6)
ITEM NO	ARMY	AIR FORCE	USMC	CAGEC	PART NUMBER	DESCRIPTION AND USABLE ON CODE (UOC)	QTY

GROUP 0602 INTAKE AND EXHAUST VALVES
FIGURE 24 CYLINDER HEAD, INTAKE, AND EXHAUST VALVES

1	XBFFF	XBFFF	XBFFF	66836	RE57234	CYLINDER HEAD,DIESE...................1	
2	XBFZZ	XB	XBFZZ	66836	RE42713	PLUG,PIPE............................1	
3	PAFZZ	PAFZZ	PAFZZ	75160	R26125	SPRING,SPECIAL...................... 8	
4	PAFZZ	PAFZZ	PAFZZ	75160	RE60005	ROTOR,ENGINE POPPET..................8	
5	PAFZZ	PAFZZ	PAFZZ	75755	R91889	LOCK,VALVE SPRING R.................16	
6	PAFZZ	PAFZZ	PAFZZ	75755	R85363	SCREW,CAP,HEXAGON H.................18	
7	XBFZZ	XB	XBFZZ	75755	R43409	CAP,PROTECTIVE,DUST..................2	
8	XBFZZ	XB	XBFZZ	75755	CD16284	CAP,PIPE............................1	
9	PAFZZ	PAFZZ	PAFZZ	75160	RE31617	SEAL,PLAIN ENCASED...................8	
10	PAFZZ	PAFZZ	PAFZZ	75755	R90692	VALVE,POPPET,ENGINE..................4	
11	PAFZZ	PAFZZ	PAFZZ	75755	R98062	VALVE,POPPET,ENGINE..................4	
12	PAFZZ	PAFZZ	PAFZZ	75755	R98063	INSERT,ENGINE VALVE..................4	
13	PAFZZ	PAFZZ	PAFZZ	66836	R116515	GASKET..............................1	

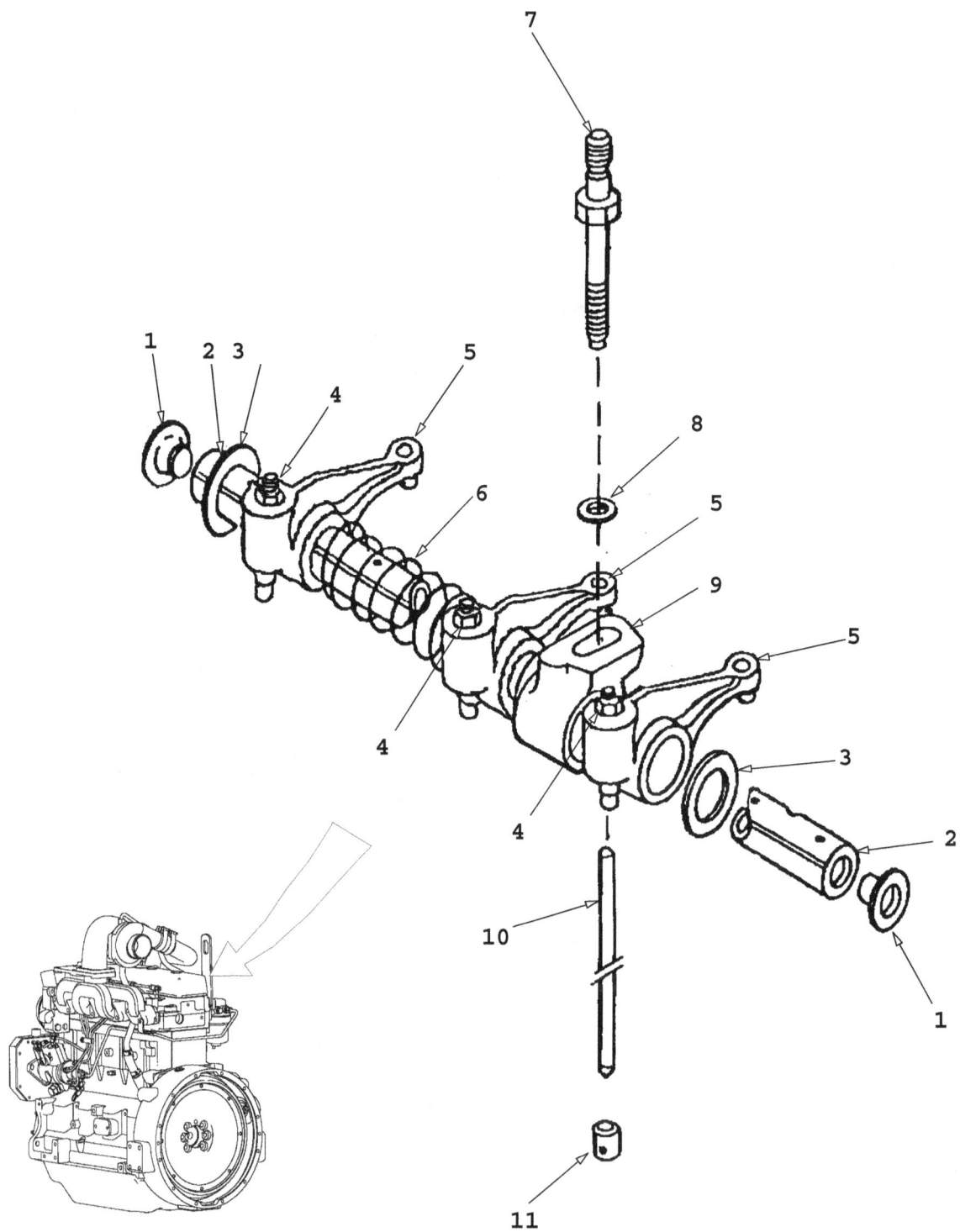

Figure 25. Rocker Arm Assembly

(1)				(2)	(3)	(4)	(5)	(6)
SMR CODE								
ITEM NO	ARMY	AIR FORCE	USMC		CAGEC	PART NUMBER	DESCRIPTION AND USABLE ON CODE (UOC)	QTY

GROUP 0603 ROCKER ARM ASSEMBLY
FIGURE 25 ROCKER ARM ASSEMBLY

1	PAFZZ	PAFZZ	PAFZZ	75160	R54565	PLUG,PROTECTIVE,DUS...................2
2	PAFZZ	PAFZZ	PAFZZ	66836	R123513	SHAFT,ROCKER ARM,EN...................1
3	PAFZZ	PAFZZ	PAFZZ	75160	T20316	WASHER,SPRING TENSI...................2
4	PAFZZ	PAFZZ	PAFZZ	66836	14M7273	NUT.......................................8
5	PAFZZ	PAFZZ	PAFZZ	66836	RE68695	ROCKER ARM,ENGINE P................. 8
6	PAFZZ	PAFZZ	PAFZZ	75160	T20314	SPRING,HELICAL,COMP...................3
7	PAFZZ	PAFZZ	PAFZZ	66836	R123271	STUD,PLAIN............................4
8	PAFZZ	PAFZZ	PAFZZ	75160	R42729	WASHER,FLAT...........................4
9	PAFZZ	PAFZZ	PAFZZ	66836	R123161	BRACKET,ROCKER ARM....................4
10	PAFZZ	PAFZZ	PAFZZ	75755	R107731	PUSH ROD,ENGINE POP...................8
11	PAFZZ	PAFZZ	PAFZZ	66836	R123565	CAM FOLLOWER,NEEDLE...................8

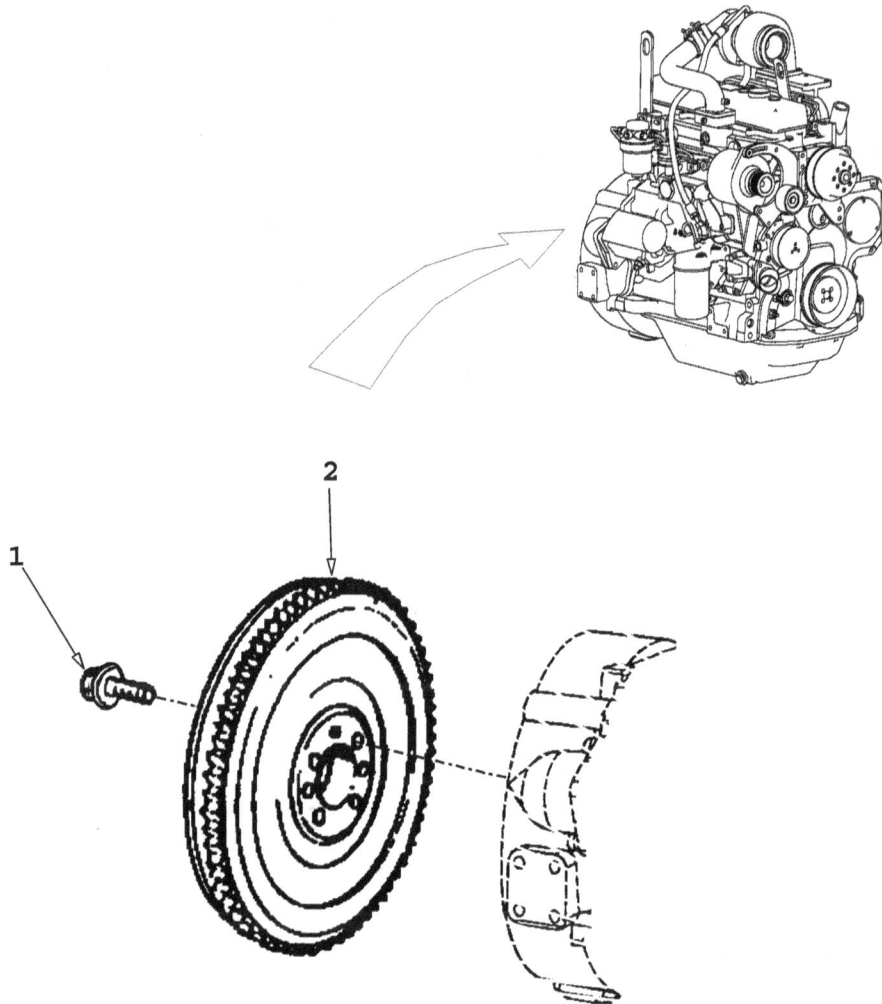

Figure 26. Flywheel

(1) SMR CODE ITEM NO	ARMY	(2) AIR FORCE	USMC	(3) CAGEC	(4) PART NUMBER	(5) DESCRIPTION AND USABLE ON CODE (UOC)	(6) QTY

GROUP 0701 FLYWHEEL
FIGURE 26 FLYWHEEL

| 1 | XBFZZ | XB | XBFZZ | 66836 | R135918 | SCREW,SPECIAL,FLANG...................4 | 2 |
| | XBFZZ | XB | XBFZZ | 66836 | RE500398 | FLYWHEEL,ENGINE ,STATICALLY BALANCED.................1 | |

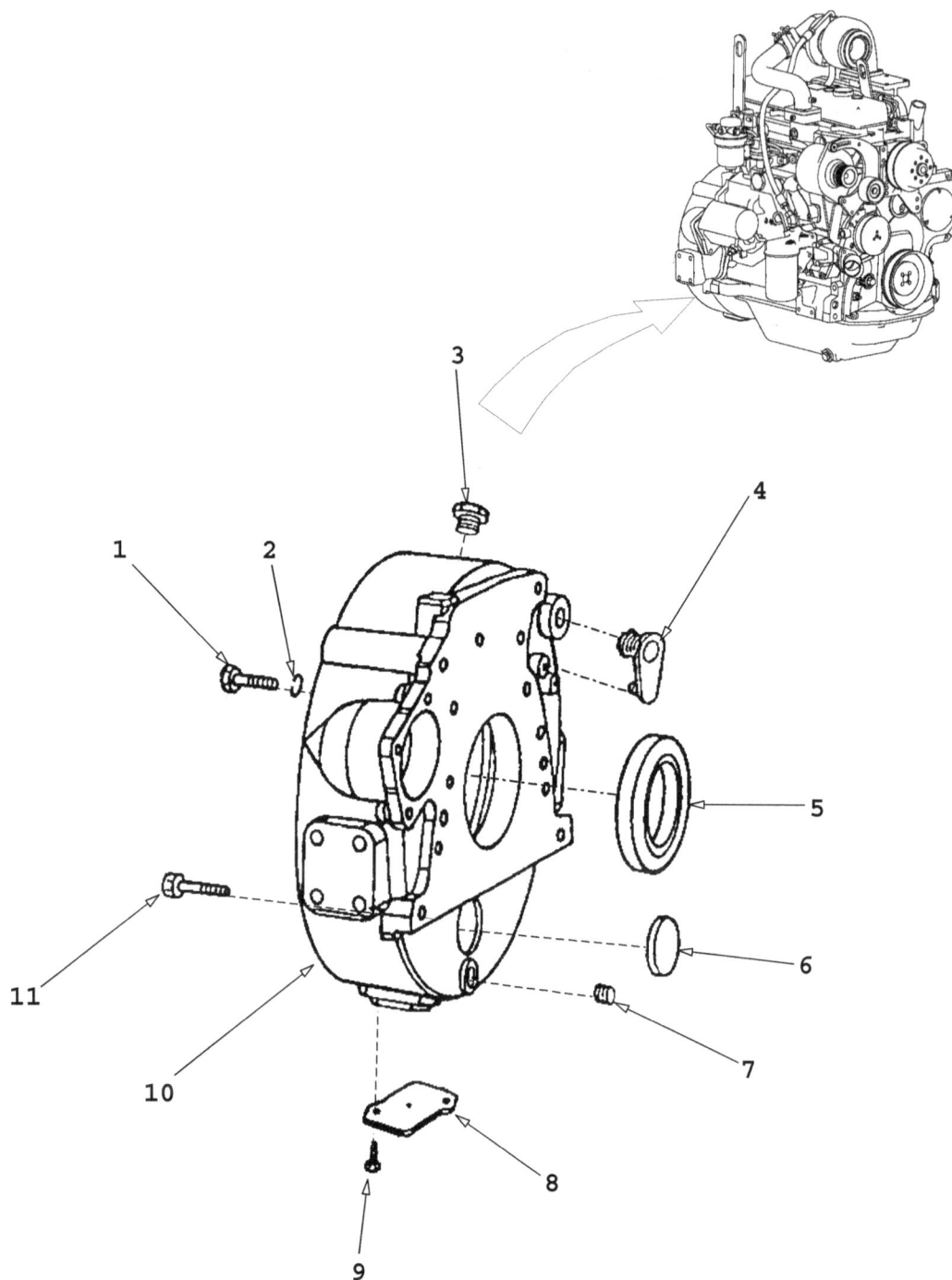

Figure 27. Flywheel Housing

(1)				(2)	(3)	(4)	(5)	(6)
SMR CODE							DESCRIPTION AND	
ITEM		AIR				PART		
NO	ARMY	FORCE	USMC		CAGEC	NUMBER	USABLE ON CODE (UOC)	QTY

GROUP 0703 FLYWHEEL HOUSING
FIGURE 27 FLYWHEEL HOUSING

1	PAFZZ	PAFZZ	PAFZZ	66836	R135918	SCREW,SPECIAL,FLANG..................8
2	PAFZZ	PAFZZ	PAFZZ	66836	T77528	WASHER,SPRING-LOCK...................8
3	XBFZZ	XB	XBFZZ	66836	R39112	PLUG,PIPE............................1
4	XBFZZ	XB	XBFZZ	66836	R131768	PLUG,MACHINE THREAD..................1
5	PAFZZ	PAFZZ	PAFZZ	66836	RE44574	SEAL,REAR OIL........................1
6	XBFZZ	XB	XBFZZ	66836	T13518	PLUG,PIPE............................1
7	XBFZZ	XB	XBFZZ	66836	15H697	FITTING,PIPE PLUG....................1
8	XBFZZ	XB	XBFZZ	66836	R121039	PLATE,COVER..........................1
9	XBFZZ	XB	XBFZZ	66836	19M7923	SCREW,CAP,HEXAGON H..................2
10	XBFZZ	XB	XBFZZ	66836	R122087	HOUSING,FLYWHEEL.....................1
11	PAFZZ	PAFZZ	PAFZZ	66836	19M8306	BOLT,MACHINE.........................1

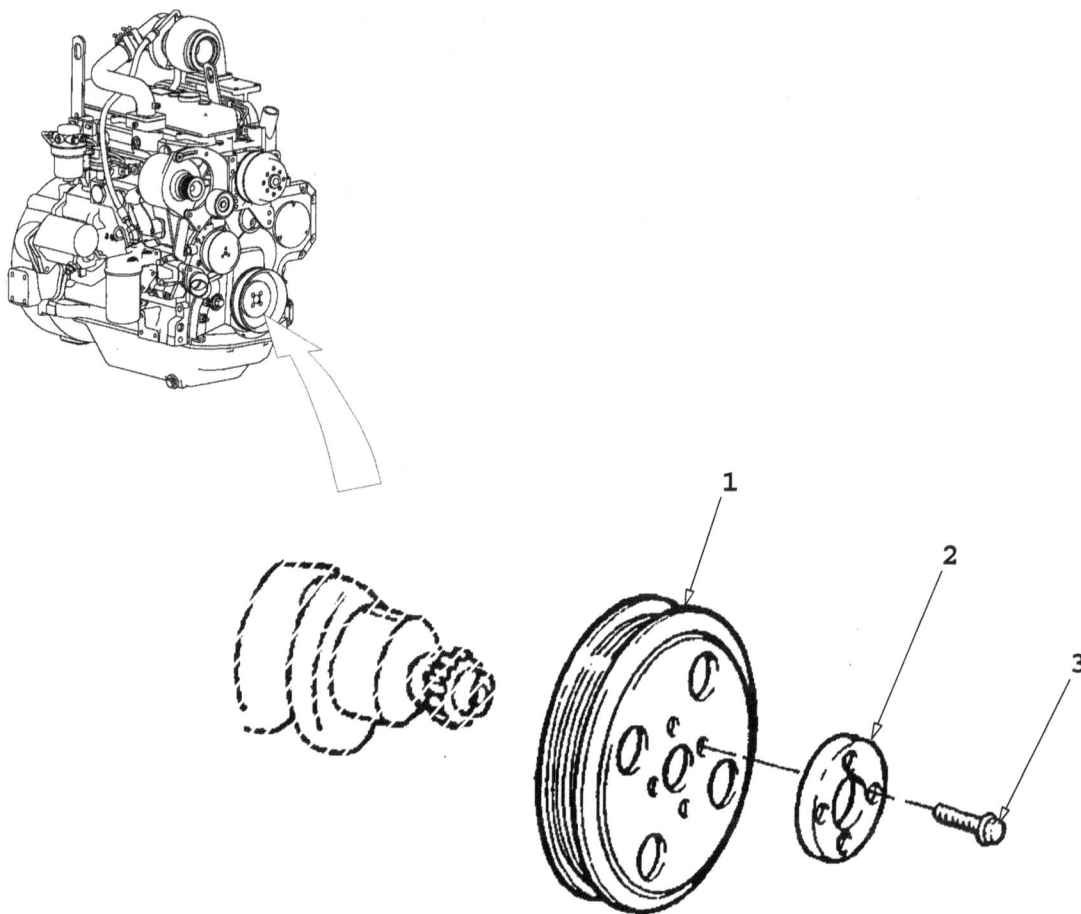

Figure 28. Crankshaft Pulley

(1)				(2)	(3)	(4)	(5)	(6)
SMR CODE							DESCRIPTION AND	
ITEM		AIR				PART		
NO	ARMY	FORCE	USMC		CAGEC	NUMBER	USABLE ON CODE (UOC)	QTY

GROUP 08 CRANKSHAFT PULLEY
FIGURE 28 CRANKSHAFT PULLEY

1	XBFZZ	XB	XBFZZ	66836	R133295	PULLEY,CRANKSHAFT.....................1		
2	XBFZZ	XB	XBFZZ	66836	R500649	FLANGE,CRANKSHAFT.....................1		
3	PAFZZ	PAFZZ	PAFZZ	66836	R121897	SCREW,CAP,HEXAGON H..................4		

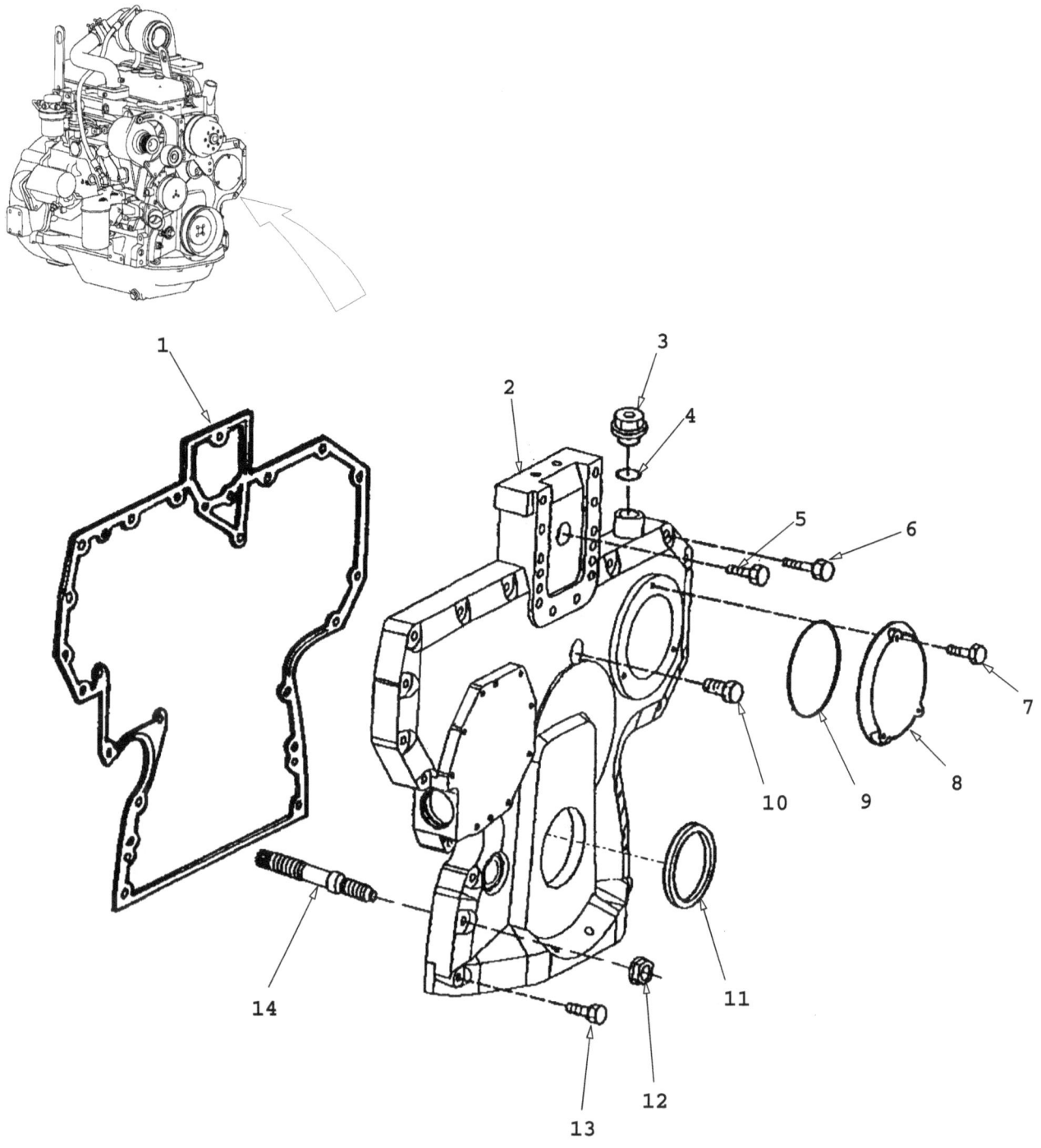

Figure 29. Timing Gear Cover

(1) SMR CODE ITEM NO	ARMY	(2) AIR FORCE	USMC	(3) CAGEC	(4) PART NUMBER	(5) DESCRIPTION AND USABLE ON CODE (UOC)	(6) QTY

GROUP 09 TIMING GEAR COVER
FIGURE 29 TIMING GEAR COVER

1	PAFZZ	PAFZZ	PAFZZ	66836	R136516	GASKET................................1
2	XBFZZ	XB	XBFZZ	66836	R134531	COVER,TIMING GEAR....................1
3	XBFZZ	XB	XBFZZ	66836	R101085	PLUG.................................1
4	PAFZZ	PAFZZ	PAFZZ	66836	51M7044	O-RING...............................1
5	PAFZZ	PAFZZ	PAFZZ	66836	19M8291	SCREW,FLANGED........................1
6	PAFZZ	PAFZZ	PAFZZ	66836	19M8317	BOLT,MACHINE.........................6
7	PAFZZ	PAFZZ	PAFZZ	75755	19M7775	BOLT,MACHINE.........................3
8	XBFZZ	XB	XBFZZ	66836	R121411	COVER,INJECTION PUM..................1
9	PAFZZ	PAFZZ	PAFZZ	66836	R121424	O-RING...............................1
10	PAFZZ	PAFZZ	PAFZZ	66836	19M7801	BOLT,MACHINE.........................1
11	PAFZZ	PAFZZ	PAFZZ	75160	RE59810	SEAL,PLAIN,FRONT OIL.................1
12	PAFZZ	PAFZZ	PAFZZ	66836	14M7296	NUT,HEX,FLANGED......................6
13	PAFZZ	PAFZZ	PAFZZ	66836	19M7979	BOLT,MACHINE.........................2
14	PAFZZ	PAFZZ	PAFZZ	66836	R123584	STUD.................................4

1

P/O 1

P/O 1

3

2

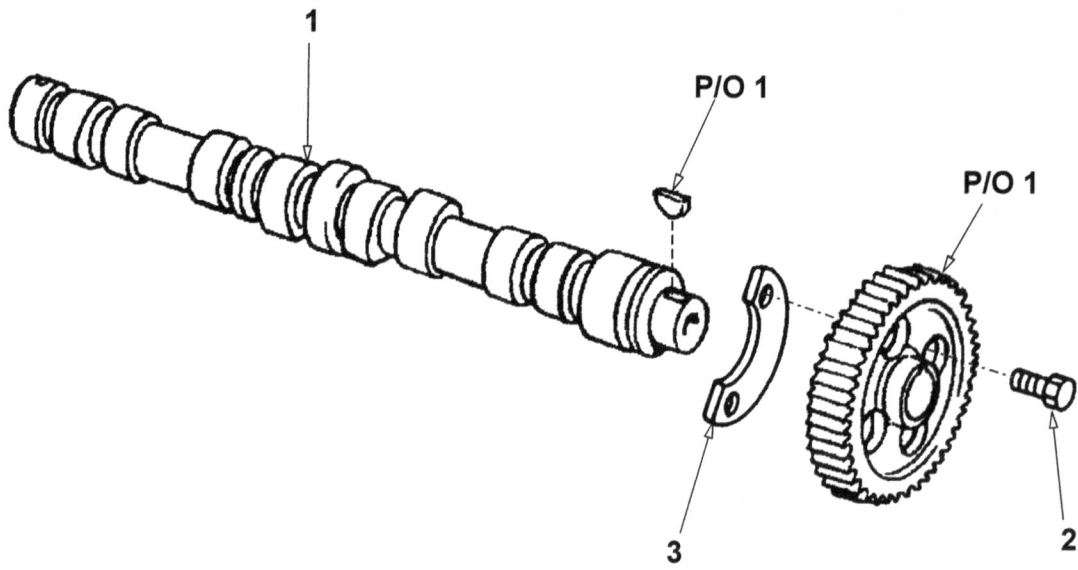

Figure 30. Camshaft Assembly

(1)				(3)	(4)	(5)	(6)
SMR CODE							
ITEM		AIR			PART	DESCRIPTION AND	
NO	ARMY	FORCE	USMC	CAGEC	NUMBER	USABLE ON CODE (UOC)	QTY

GROUP 1001 CAMSHAFT ASSEMBLY
FIGURE 30 CAMSHAFT ASSEMBLY

1	XBHZZ	XB	XBHZZ	66836	RE56375	CAMSHAFT,ENGINE.......................1	
2	PAHZZ	PAFZZ	PAHZZ	75755	19M7867	BOLT,MACHINE..........................2	
3	PAFZZ	PAFZZ	PAFZZ	66836	R132518	PLATE,THRUST..........................1	

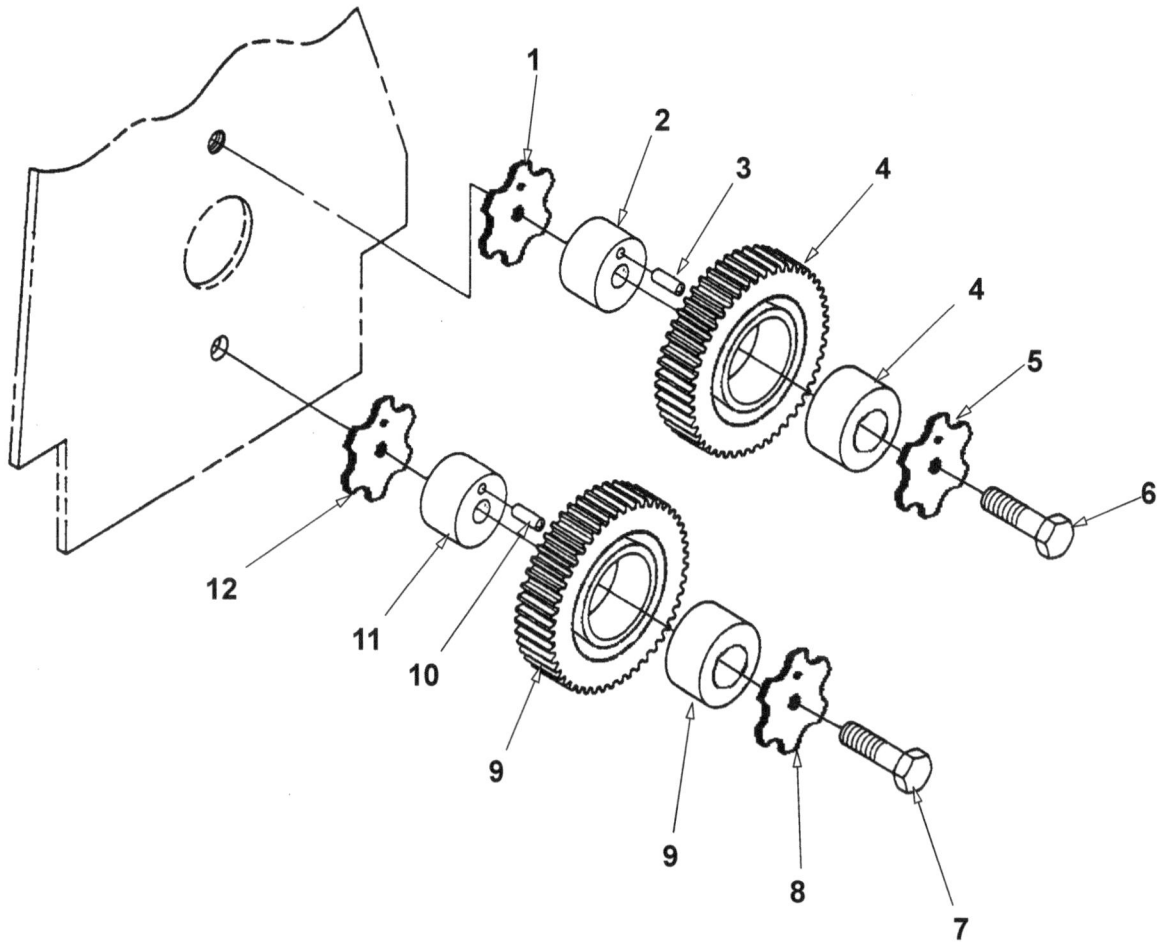

Figure 31. Idler Gears and Idler Gear Shafts

(1) SMR CODE			(3)	(4)	(5)	(6)
ITEM NO	ARMY	AIR FORCE	USMC	CAGEC	PART NUMBER	DESCRIPTION AND USABLE ON CODE (UOC)	QTY

GROUP 1002 IDLER GEARS AND IDLER GEAR SHAFTS
FIGURE 31 IDLER GEARS AND IDLER GEAR SHAFTS

Item	Army	Air Force	USMC	CAGEC	Part Number	Description	QTY
1	PAHZZ	PAFZZ	PAHZZ	66836	R123174	WASHER,THRUST..........................1
2	PAHZZ	PAFZZ	PAHZZ	66836	R120641	SHAFT,SPUR GEAR,IDLER................1
3	PAHZZ	PAFZZ	PAHZZ	75160	34H288	PIN,SPRING...........................1
4	XBHZZ	XB	XBHZZ	66836	RE56313	GEAR,SPUR,UPPER IDLER................1
5	PAHZZ	PAFZZ	PAHZZ	66836	R131206	BEARING,WASHER,THRU..................1
6	PAHZZ	PAFZZ	PAHZZ	75755	19M7808	BOLT,MACHINE.........................1
7	PAHZZ	PAFZZ	PAHZZ	66836	19M8292	BOLT,MACHINE.........................1
8	PAHZZ	PAFZZ	PAHZZ	66836	R131283	BEARING,WASHER,THRU..................1
9	XBHZZ	XB	XBHZZ	66836	RE56369	GEAR,SPUR,LOWER IDLER................1
10	PAHZZ	PAFZZ	PAHZZ	75160	34H283	PIN,SPRING...........................1
11	PAHZZ	PAFZZ	PAHZZ	66836	R114194	SHAFT,SPUR GEAR,IDLER................1
12	PAHZZ	PAFZZ	PAHZZ	66836	R101225	WASHER,THRUST........................1

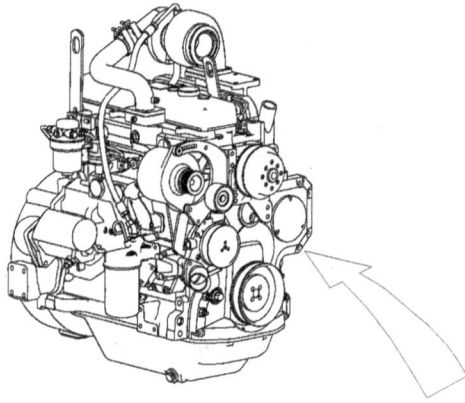

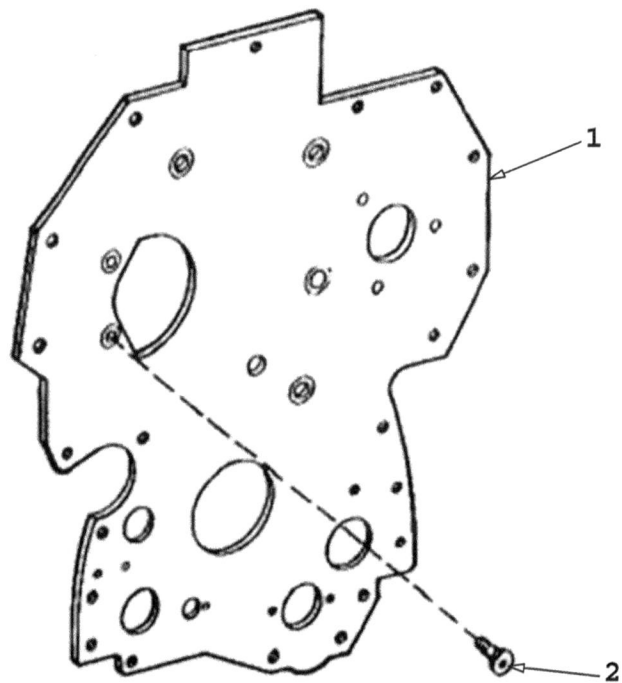

Figure 32. Front Plate

(1)		(2)		(3)	(4)	(5)	(6)
SMR CODE							
ITEM		AIR			PART	DESCRIPTION AND	
NO	ARMY	FORCE	USMC	CAGEC	NUMBER	USABLE ON CODE (UOC)	QTY

GROUP 1003 FRONT PLATE
FIGURE 32 FRONT PLATE

| 1 | XBHZZ | XB | XBHZZ | 66836 | R134527 | PLATE,METAL,CYLINDER BLOCK...........1 |
| 2 | PAHZZ | PAFZZ | PAHZZ | 66836 | R136475 | SCREW,CAP...........................6 |

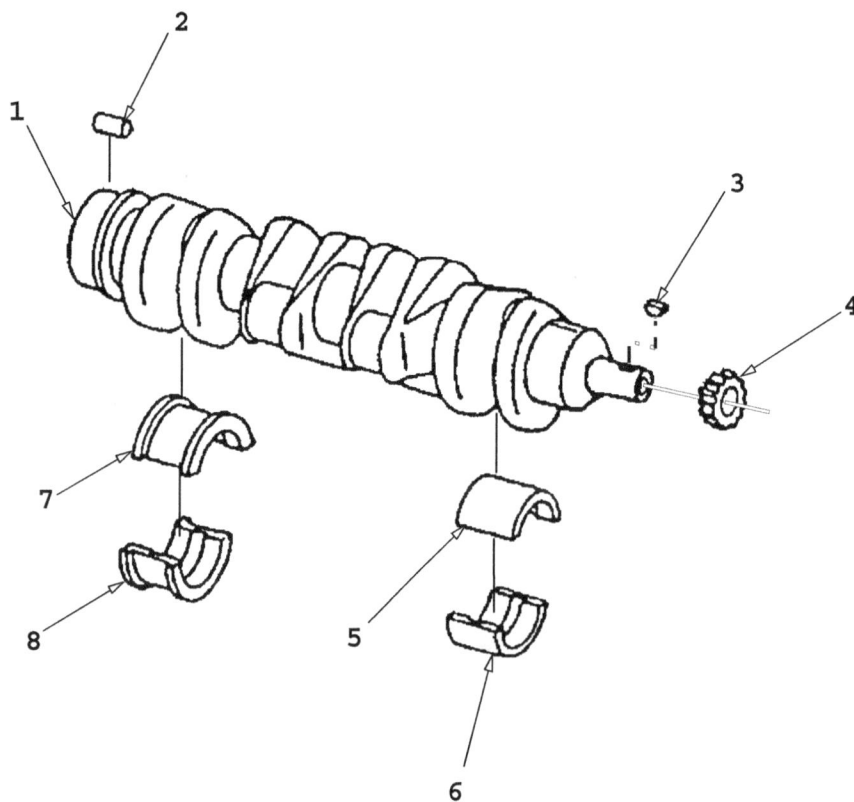

Figure 33. Crankshaft and Main Bearings

(1) SMR CODE			(2)	(3)	(4)	(5)	(6)
ITEM NO	ARMY	AIR FORCE	USMC	CAGEC	PART NUMBER	DESCRIPTION AND USABLE ON CODE (UOC)	QTY

GROUP 1004 CRANKSAHFT AND MAIN BEARINGS
FIGURE 33 CRANKSHAFT AND MAIN BEARINGS

1	XBHZZ	XB	XBHZZ	66836	RE50618	CRANKSHAFT,ENGINE....................1	
2	XBHZZ	XB	XBHZZ	00141	A120	PIN,STRAIGHT,HEADLE.................1	
3	PAHZZ	PAFZZ	PAHZZ	66836	26M4224	KEY,WOODRUFF.........................1	
4	PAHZZ	PAFZZ	PAHZZ	66836	R120631	GEAR,SPUR............................1	
5	PAHZZ	PAFZZ	PAHZZ	66836	R123561	BEARING,MAIN,UPPER...................4	
6	PAHZZ	PAFZZ	PAHZZ	66836	R123562	BEARING,MAIN,LOWER...................4	
7	PAHZZ	PAFZZ	PAHZZ	66836	R123563	BEARING,WASHER,THRU.................1	
8	PAHZZ	PAFZZ	PAHZZ	66836	R123564	BEARING,WASHER,THRU.................1	

P/O 1

P/O 1

1

2

7

2

6

3

3

3

4

5

P/O
4

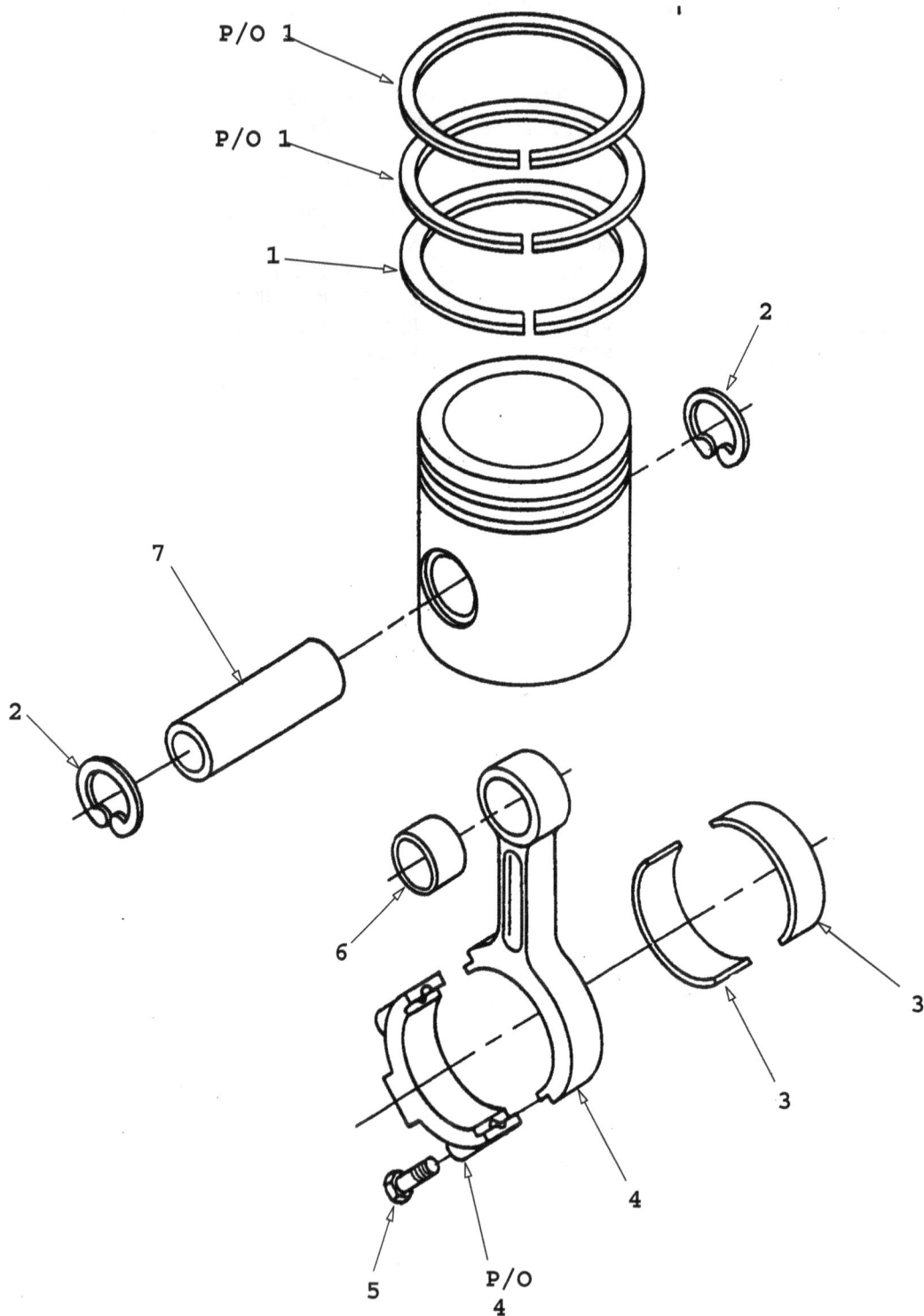

Figure 34.   Pistons and Connecting Rods

(1)		(2)		(3)	(4)	(5)	(6)
SMR CODE							
ITEM		AIR			PART	DESCRIPTION AND	
NO	ARMY	FORCE	USMC	CAGEC	NUMBER	USABLE ON CODE (UOC)	QTY

GROUP 1005 PISTONS AND CONNECTING RODS
FIGURE 34 PISTONS AND CONNECTING RODS

1	PAHZZ	PAFZZ	PAHZZ	75160	RE66271	KIT,PISTON RING......................1	
2	PAHZZ	PAFZZ	PAHZZ	96906	MS16625-1137	RING,RETAINING......................2	
3	PAHZZ	PAFZZ	PAHZZ	66836	R113698	BEARING HALF........................8	
4	XBHZZ	XB	XBHZZ	66836	RE60272	CONNECTING ROD,PIST.................1	
5	PAHZZ	PAFZZ	PAHZZ	66836	R500500	SCREW,CAP...........................2	
6	PAHZZ	PAFZZ	PAHZZ	66836	R123960	BUSHING.............................1	
7	PAHZZ	PAFZZ	PAHZZ	66836	R123178	PIN,PISTON..........................1	

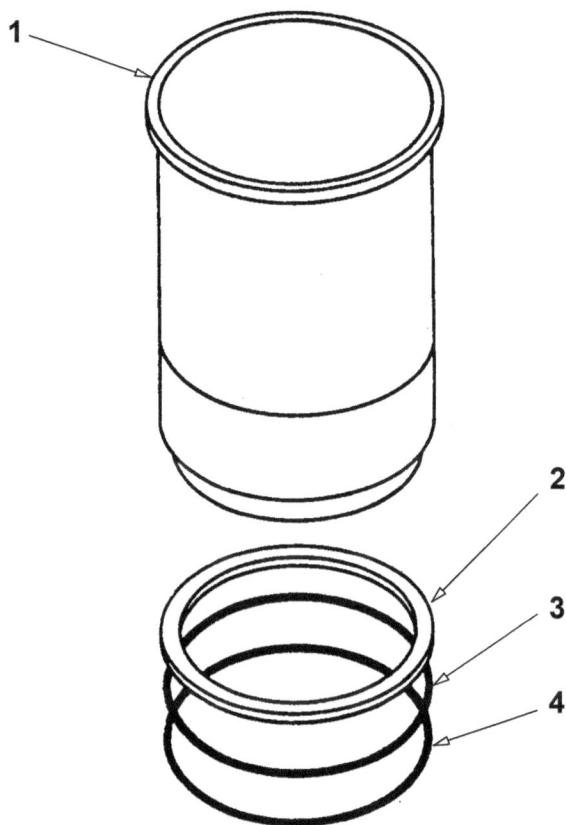

Figure 35. Cylinder Liner

(1)			(3)	(4)	(5)	(6)
SMR CODE						
ITEM		AIR		PART	DESCRIPTION AND	
NO	ARMY	FORCE	USMC	CAGEC	NUMBER	USABLE ON CODE (UOC)	QTY

GROUP 1006 CYLINDER LINER
FIGURE 35 CYLINDER LINER

1	XBHZZ	XB	XBHZZ	66836	RE65967	KIT,PISTON LINER.....................1
2	PAHZZ	PAFZZ	PAHZZ	75160	R48767	PACKING.............................1
3	PAHZZ	PAFZZ	PAHZZ	66836	R26121	O-RING..............................1
4	PAHZZ	PAFZZ	PAHZZ	75160	R55453	O-RING..............................1

P/O 1 P/O 1

P/O 1

1

2

P/O 3

P/O

3

2

P/O 3

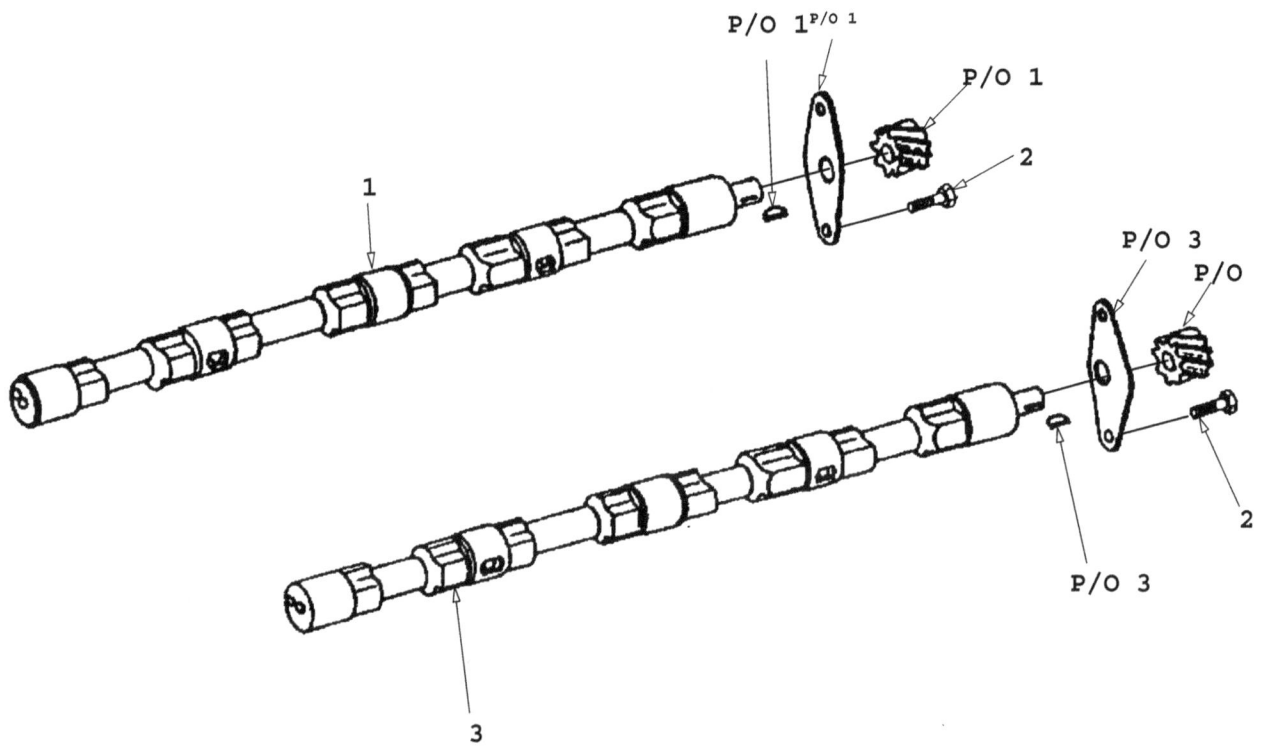

Figure 36. Balancer Shafts

(1)		(2)		(3)	(4)	(5)	(6)
SMR CODE							
ITEM		AIR			PART	DESCRIPTION AND	
NO	ARMY	FORCE	USMC	CAGEC	NUMBER	USABLE ON CODE (UOC)	QTY

GROUP 1007 BALANCER SHAFTS
FIGURE 36 BALANCER SHAFTS

1	XBHZZ	XB	XBHZZ	66836	RE69365	SHAFT,STRAIGHT,LH BALANCER...........1
2	PAHZZ	PAFZZ	PAHZZ	66836	19M7864	BOLT,MACHINE........................4
3	XBHZZ	XB	XBHZZ	66836	RE69364	SHAFT,STRAIGHT,RH BALANCER...........1

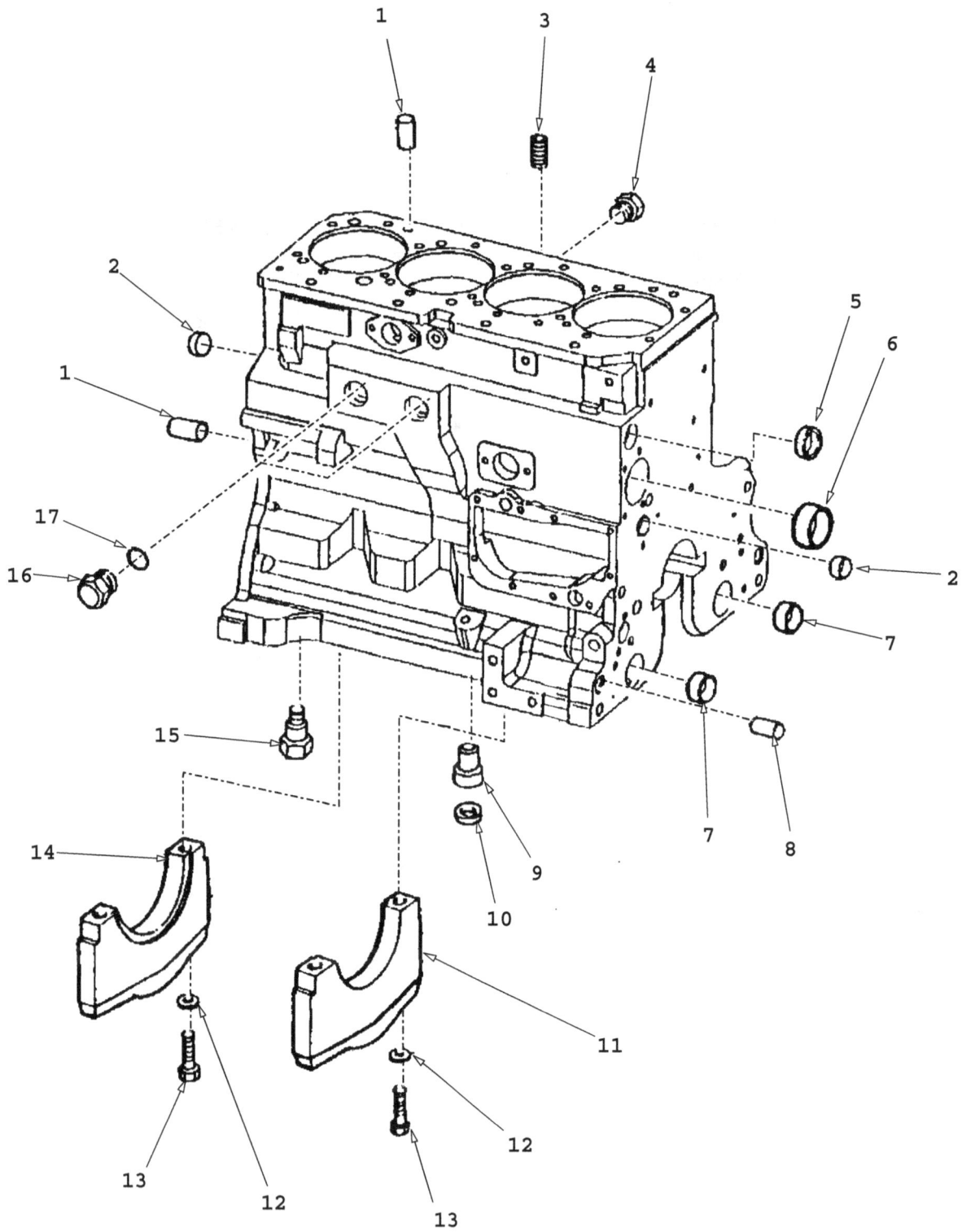

Figure 37. Cylinder Block

(1)	(2)			(3)	(4)	(5)	(6)
SMR CODE							
ITEM		AIR			PART	DESCRIPTION AND	
NO	ARMY	FORCE	USMC	CAGEC	NUMBER	USABLE ON CODE (UOC)	QTY

GROUP 1107 CYLINDER BLOCK
FIGURE 37 CYLINDER BLOCK

1	XBHZZ	XB	XBHZZ	66836	B153	DOWEL..............................2	
2	XBHZZ	XB	XBHZZ	75755	T18891	PLUG...............................2	
3	XBHZZ	XB	XBHZZ	75160	15H697	FITTING,PIPE PLUG..................1	
4	XBHZZ	XB	XBHZZ	66836	RE42713	PLUG,PIPE..........................5	
5	XBHZZ	XB	XBHZZ	66836	R116466	PLUG...............................1	
6	PAHZZ	PAFZZ	PAHZZ	66836	R119874	BUSHING,SLEEVE.....................1	
7	PAHZZ	PAFZZ	PAHZZ	66836	R115299	BUSHING,SLEEVE.....................6	
8	XBHZZ	XB	XBHZZ	00141	A120	PIN,STRAIGHT,HEADLE................2	
9	XBHZZ	XB	XBHZZ	66836	R115390	CONNECTOR,TUBE.....................1	
10	PAFZZ	PAFZZ	PAFZZ	75160	R75892	O-RING.............................1	11
	XBHZZ	XB	XBHZZ	66836	R114240	CAP,PILLOW BLOCK	
						,MAIN BEARING.......................1	
12	PAHZZ	PAFZZ	PAHZZ	75160	T20168	WASHER,FLAT.......................10	
13	PAHZZ	PAFZZ	PAHZZ	75160	T23474	SCREW,CAP,HEXAGON H...............10	
14	XBHZZ	XB	XBHZZ	66836	R114241	CAP,PILLOW BLOCK,BEARING...........4	
15	PAHZZ	PAFZZ	PAHZZ	66836	R131182	RESTRICTOR,FLUID FL................4	
16	XBHZZ	XB	XBHZZ	75160	R39741	PLUG,MACHINE THREAD................2	
17	PAHZZ	PAFZZ	PAHZZ	06021	U13639	O-RING.............................2	

P/O
1

1

2

1

3

P/O
3

P/O
3

P/O
3

4

P/O
3

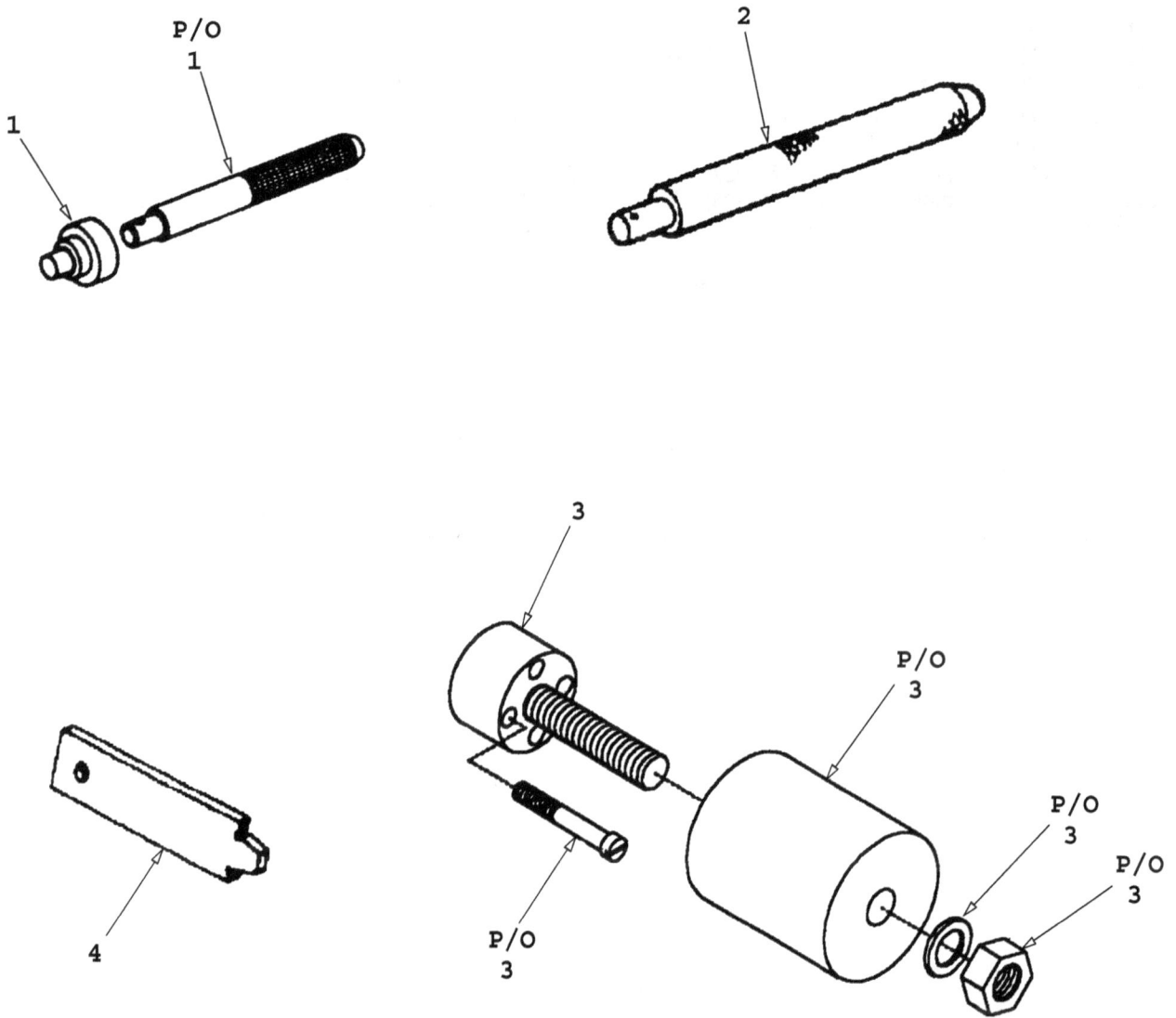

Figure 38.   Special Tools and test Equipment

(1) SMR CODE			(2)	(3)	(4)	(5)	(6)
ITEM NO	ARMY	AIR FORCE	USMC	CAGEC	PART NUMBER	DESCRIPTION AND USABLE ON CODE (UOC)	QTY

GROUP 12 SPECIAL TOOLS AND TEST EQUIPMENT
FIGURE 38 SPECIAL TOOLS AND TEST EQUIPMENT

1	PEFZZ	PEFZZ	PEFZZ	66836	JD248A	INSERTER,BEARING BU..................1	
2	PEFZZ	PEFZZ	PEFZZ	66836	JDG536	HANDLE,BUSHING DRIV..................1	
3	PEFZZ	PEFZZ	PEFZZ	66836	JDG954	SET,OIL SEAL INSTAL..................1	
4	PEHZZ	PEFZZ	PEFZZ	66836	JDE62	GAUGE,WEAR..........................1	

NATIONAL STOCK NUMBER INDEX

STOCK NUMBER	FIG.	ITEM	STOCK NUMBER	FIG.	ITEM
5330-00-007-3105	35	2	5306-01-330-8490	30	2
5360-00-009-9260	24	3	5310-01-331-7601	1	26
3110-00-100-2357	5	3		18	10
3110-00-100-2364	5	1	2815-01-333-0730	24	12
5310-00-350-4717	5	22	2815-01-333-1457	24	10
5365-00-754-1083	34	2	2910-01-333-2238	12	1
3120-01-106-2405	15	1	5310-01-333-2754	1	7
5365-01-118-4113	22	3	2815-01-335-3545	24	5
5305-01-119-1059	37	13	2815-01-335-4414	24	11
5360-01-119-1609	25	6	5306-01-337-1978	1	10
5315-01-119-3116	31	10		2	8
5310-01-119-3672	37	12	2920-01-359-4770	1	9
5310-01-119-4371	25	8	5310-01-360-1690	5	14
5340-01-131-0078	25	1	2920-01-360-7746	5	2
5310-01-134-3292	25	3	5330-01-360-9021	24	9
4720-01-195-3850	8	2	5330-01-360-9028	5	15
5310-01-197-2389	5	23	5310-01-360-9083	5	13
5315-01-197-3144	31	3	3110-01-361-0251	5	27
5365-01-197-4895	5	31	2920-01-361-2361	5	4
4730-01-232-2136	22	4	5310-01-361-2649	5	20
5340-01-235-3309	20	2	5306-01-361-5612	5	12
5310-01-255-2619	1	18	5305-01-361-6742	5	21
5310-01-257-3234	1	16	5310-01-364-4303	14	3
5330-01-257-9152	37	17	5360-01-374-8301	5	7
5330-01-277-6141	35	4	5360-01-381-6121	5	26
5330-01-292-2440	9	2	5340-01-383-8775	20	11
5330-01-307-9409	27	5	2815-01-395-8859	25	10
5330-01-307-9411	37	10	4730-01-408-1575	8	1
5330-01-307-9424	8	6	5360-01-408-5368	15	2
5331-01-310-8449	7	6	5310-01-408-5853	1	1
5330-01-311-5972	14	6	5331-01-409-1634	3	7
4730-01-318-9677	20	15	5310-01-423-5301	1	6
5306-01-326-4911	7	10	3110-01-425-9914	5	17
	10	6	5340-01-425-9915	5	19
	11	10	3020-01-435-8532	5	16
5306-01-326-4912	29	7	5330-01-444-0829	13	1
5310-01-328-7657	1	5	2815-01-444-1694	34	1
	25	4	2815-01-444-2323	24	4
4730-01-328-8545	1	27	4330-01-444-3729	10	4
	29	12	2910-01-444-3758	18	7
5305-01-329-6988	24	6	6685-01-444-9477	2	4
5305-01-329-9729	1	12	5330-01-452-0929	18	8
	7	5	5330-01-452-1310	18	1
	8	5	4720-01-460-2609	7	4
	18	5	5331-01-460-2664	23	2
5306-01-330-0426	3	4	5331-01-460-2668	13	3
5306-01-330-5196	31	6	5330-01-460-2670	7	11
5306-01-330-5197	1	14	5330-01-460-2677	3	1
5306-01-330-8490	1	17	2910-01-460-8973	22	1
	3	5	2950-01-461-1377	7	1

NATIONAL STOCK NUMBER INDEX

STOCK NUMBER	FIG.	ITEM	STOCK NUMBER	FIG.	ITEM
2920-01-469-9250	4	5	5306-01-470-1993	10	7
4720-01-470-0320	20	7		11	12
4720-01-470-0337	20	10		14	8
4720-01-470-0339	20	9	5306-01-470-1994	13	2
4720-01-470-0343	20	8	5330-01-470-2010	29	4
5310-01-470-0445	1	2	5330-01-470-2014	14	5
4720-01-470-0447	20	1	5330-01-470-2019	35	3
5310-01-470-0459	23	3	5330-01-470-2022	16	5
5310-01-470-0466	1	11	5330-01-470-2025	29	9
5310-01-470-0551	31	12	5330-01-470-2027	2	10
5310-01-470-0563	31	1	5330-01-470-2034	2	3
5310-01-470-0643	19	4	5330-01-470-2067	19	2
5310-01-470-0658	12	3	5305-01-470-2417	29	5
4710-01-470-0678	20	4	5305-01-470-2418	34	5
4710-01-470-0679	2	12	5305-01-470-2422	10	5
	10	1		11	11
	11	13	5305-01-470-2424	19	5
4730-01-470-0685	2	1	5340-01-470-2539	4	10
4730-01-470-0735	7	13	4730-01-470-3059	7	2
5340-01-470-0880	23	4	4730-01-470-3060	19	3
4730-01-470-0897	7	15	3120-01-470-3553	34	3
4730-01-470-1288	7	7	3110-01-470-3555	33	8
4710-01-470-1297	7	12	3120-01-470-3571	34	6
5306-01-470-1335	1	22	3120-01-470-3574	6	5
5340-01-470-1357	18	11	3110-01-470-3581	25	11
5306-01-470-1372	31	7	3110-01-470-3585	33	7
5306-01-470-1375	36	2	3120-01-470-3588	33	6
5306-01-470-1379	2	13	3120-01-470-3591	33	5
	29	10	3110-01-470-3594	30	3
5306-01-470-1384	18	13	5305-01-470-3595	32	3
5306-01-470-1389	29	6	5306-01-470-3597	4	1
5306-01-470-1391	3	9	3120-01-470-3598	37	6
5306-01-470-1395	3	8	3120-01-470-3600	37	7
5305-01-470-1585	5	30	3120-01-470-3606	31	8
5305-01-470-1599	5	11	3120-01-470-3610	31	5
5365-01-470-1621	18	2	3110-01-470-3616	4	4
	20	13	3110-01-470-3618	4	7
2915-01-470-1640	18	6	5360-01-470-3626	12	2
5365-01-470-1688	31	11	4820-01-470-3846	15	3
5365-01-470-1695	31	2	3020-01-470-3853	33	4
5330-01-470-1730	29	11	2815-01-470-3857	25	5
5306-01-470-1945	2	6	3110-01-470-3864	5	3
5365-01-470-1948	6	3	3030-01-470-3867	6	7
5306-01-470-1950	9	3	3110-01-470-3870	5	3
5306-01-470-1959	2	2	2930-01-470-3895	16	1
5306-01-470-1960	13	5	6680-01-470-3952	17	1
5306-01-470-1961	28	3	2930-01-470-4157	3	2
5306-01-470-1964	27	1	5315-01-470-4170	33	3
5306-01-470-1984	27	9	4730-01-470-4178	37	15
5306-01-470-1986	22	5	2910-01-470-4183	19	1

NATIONAL STOCK NUMBER INDEX

STOCK NUMBER	FIG.	ITEM	STOCK NUMBER	FIG.	ITEM
5306-01-470-4189	16	3			
5306-01-470-4193	16	2			
	29	13			
5306-01-470-4194	27	11			
5330-01-470-4200	29	1			
2815-01-470-4202	34	7			
2815-01-470-4214	25	9			
5307-01-470-4220	1	8			
5307-01-470-4221	29	14			
	32	4			
5307-01-470-4223	32	1			
5307-01-470-4225	25	7			
5977-01-470-4249	5	6			
2920-01-470-4265	5	25			
3020-01-470-4287	5	32			
5306-01-470-5067	1	23			
5305-01-470-5089	1	19			
5306-01-470-5111	1	25			
5306-01-470-5118	6	6			
5330-01-470-5139	10	2			
5330-01-470-5140	7	9			
5331-01-470-5159	7	8			
5331-01-470-5162	7	3			
5330-01-470-5183	11	3			
5330-01-470-5305	24	13			
5330-01-470-5475	11	6			
5330-01-470-5500	11	7			
5330-01-470-5524	11	2			
5306-01-470-6008	11	5			
5306-01-470-6013	6	2			
2815-01-470-6283	14	1			
2930-01-470-6284	11	4			
3020-01-470-6328	6	4			
2815-01-470-6375	25	2			
6115-01-470-6402	4	2			
5977-01-470-6406	4	16			
6110-01-470-6410	4	13			
2920-01-470-7383	4	9			
3020-01-470-7520	14	2			
5330-01-470-7663	38	3			
4730-01-471-1512	20	5			
5331-01-471-7464	20	14			
5210-01-474-6543	38	4			

PART NUMBER INDEX

CAGEC	PART NUMBER	STOCK NUMBER	FIG.	ITEM
75755	AR85519		20	6
92836	AT175195		1	29
66836	A120		33	2
			37	8
66836	A4827	5310-01-470-0658	12	3
66836	B153		37	1
75755	CD16284		24	8
53867	GB610	5977-01-470-6406	4	16
66836	H23125	5340-01-470-0880	23	4
66836	JDE62	5210-01-474-6543	38	4
66836	JDG536		38	2
66836	JDG954	5330-01-470-7663	38	3
66836	JD248A		38	1
96906	MS16625-1137	5365-00-754-1083	34	2
75160	RE19797	5340-01-235-3309	20	2
75160	RE19799	5340-01-383-8775	20	11
66836	RE26409		7	16
75160	RE31617	5330-01-360-9021	24	9
66836	RE42713		24	2
			37	4
75160	RE44574	5330-01-307-9409	27	5
66836	RE46684		2	7
66836	RE46685		11	14
66836	RE46686		11	15
			18	4
75160	RE48786	2910-01-460-8973	22	1
75755	RE48827	2920-01-359-4770	1	9
66836	RE500398		26	2
66836	RE500539		6	9
66836	RE500737	2930-01-470-4157	3	2
66836	RE501022		13	6
66836	RE501453		3	6
66836	RE501472	2930-01-470-3895	16	1
66836	RE502269	2815-01-470-6283	14	1
66836	RE505411		1	4
66836	RE50618		33	1
75755	RE50752	5330-01-452-0929	18	8
75755	RE51649	5330-01-452-1310	18	1
66836	RE54885	4730-01-470-0897	7	15
66836	RE55871	4730-01-470-3059	7	2
66836	RE56313		31	4
66836	RE56369		31	9
66836	RE56375		30	1
66836	RE57234		24	1
66836	RE57619		14	7
66836	RE59341		8	3
66836	RE59447	4720-01-470-0337	20	10
66836	RE59448	4720-01-470-0339	20	9
66836	RE59449	4720-01-470-0343	20	8
66836	RE59450	4720-01-470-0320	20	7
75160	RE59468	4720-01-460-2609	7	4

PART NUMBER INDEX

CAGEC	PART NUMBER	STOCK NUMBER	FIG.	ITEM
66836	RE59547	4710-01-470-1297	7	12
66836	RE59801		23	6
75160	RE59810	5330-01-470-1730	29	11
66836	RE59812	2930-01-470-6284	11	4
66836	RE59907		10	3
75160	RE60005	2815-01-444-2323	24	4
75160	RE60021	2910-01-444-3758	18	7
66836	RE60022		18	14
66836	RE60025		18	9
66836	RE60029	5365-01-470-1621	18	2
			20	13
66836	RE60037		18	3
66836	RE60071		8	4
75160	RE60076	2950-01-461-1377	7	1
66836	RE60272		34	4
66836	RE60854	2915-01-470-1640	18	6
75160	RE64354	6685-01-444-9477	2	4
66836	RE65908	3120-01-470-3553	34	3
66836	RE65967		35	1
66836	RE65978	4730-01-470-0735	7	13
75160	RE66271	2815-01-444-1694	34	1
66836	RE66298	5305-01-470-2424	19	5
66836	RE68345	2910-01-470-4183	19	1
66836	RE68695	2815-01-470-3857	25	5
66836	RE68722	3020-01-470-6328	6	4
66836	RE68748	4710-01-470-0678	20	4
66836	RE69243	6680-01-470-3952	17	1
66836	RE69364		36	3
66836	RE69365		36	1
66836	RE70401		23	1
66836	RE70848	4720-01-470-0447	20	1
66836	RE71526	5306-01-470-4189	16	3
66836	R101085		29	3
66836	R101225	5310-01-470-0551	31	12
75755	R107731	2815-01-395-8859	25	10
66836	R107749	3120-01-470-3574	6	5
66836	R107770		13	4
75160	R110531	4730-01-408-1575	8	1
75160	R111137	5360-01-408-5368	15	2
66836	R114194	5365-01-470-1688	31	11
66836	R114240		37	11
66836	R114241		37	14
66836	R115138		7	14
66836	R115252		11	8
66836	R115299	3120-01-470-3600	37	7
66836	R115390		37	9
66836	R116116		18	12
66836	R116324		23	5
66836	R116366	5306-01-470-1986	22	5
66836	R116386	5307-01-470-4220	1	8
66836	R116466		37	5

PART NUMBER INDEX

CAGEC	PART NUMBER	STOCK NUMBER	FIG.	ITEM
66836	R116515	5330-01-470-5305	24	13
66836	R119874	3120-01-470-3598	37	6
66836	R120631	3020-01-470-3853	33	4
66836	R120638	3020-01-470-7520	14	2
66836	R120641	5365-01-470-1695	31	2
66836	R121038		3	3
66836	R121039		27	8
66836	R121043	4820-01-470-3846	15	3
66836	R121376		14	4
66836	R121411		29	8
66836	R121424	5330-01-470-2025	29	9
66836	R121897	5306-01-470-1961	28	3
66836	R122087		27	10
66836	R123161	2815-01-470-4214	25	9
66836	R123174	5310-01-470-0563	31	1
66836	R123178	2815-01-470-4202	34	7
66836	R123226	5330-01-470-2027	2	10
66836	R123271	5307-01-470-4225	25	7
66836	R123323	4730-01-470-0685	2	1
66836	R123324	4710-01-470-0679	2	12
			10	1
			11	13
75160	R123352	5330-01-444-0829	13	1
75160	R123417	5330-01-460-2677	3	1
66836	R123426		2	5
66836	R123471		11	1
66836	R123501	5330-01-470-5183	11	3
66836	R123502	5330-01-470-5475	11	6
66836	R123513	2815-01-470-6375	25	2
66836	R123525	5330-01-470-5139	10	2
66836	R123561	3120-01-470-3591	33	5
66836	R123562	3120-01-470-3588	33	6
66836	R123563	3110-01-470-3585	33	7
66836	R123564	3110-01-470-3555	33	8
66836	R123565	3110-01-470-3581	25	11
75160	R123570	5330-01-460-2670	7	11
66836	R123572	5330-01-470-5140	7	9
66836	R123574	5310-01-470-0459	23	3
75160	R123575	5331-01-460-2664	23	2
66836	R123584	5307-01-470-4221	29	14
			32	4
66836	R123593	4730-01-471-1512	20	5
66836	R123960	3120-01-470-3571	34	6
66836	R128658		6	1
66836	R131182	4730-01-470-4178	37	15
66836	R131206	3120-01-470-3610	31	5
66836	R131283	3120-01-470-3606	31	8
66836	R131425	5360-01-470-3626	12	2
66836	R131768		27	4
66836	R132267		1	3
66836	R132373		9	1

PART NUMBER INDEX

PART NUMBER INDEX

CAGEC	PART NUMBER	STOCK NUMBER	FIG.	ITEM
75755	R85363	5305-01-329-6988	24	6
75160	R89944	5331-01-409-1634	3	7
75160	R90658	5330-01-292-2440	9	2
75755	R90692	2815-01-333-1457	24	10
75160	R91360	5310-01-408-5853	1	1
75755	R91692		12	4
75755	R91889	2815-01-335-3545	24	5
66836	R94350		1	20
66836	R97185	5330-01-470-2014	14	5
75755	R98062	2815-01-335-4414	24	11
75755	R98063	2815-01-333-0730	24	12
66836	T122075	5330-01-470-5524	11	2
66836	T13518		27	6
66836	T14050	5340-01-470-1357	18	11
66836	T158584		1	21
75755	T18891		37	2
75160	T20168	5310-01-119-3672	37	12
75160	T20314	5360-01-119-1609	25	6
75160	T20316	5310-01-134-3292	25	3
75160	T23474	5305-01-119-1059	37	13
75755	T27860	5310-01-333-2754	1	7
66836	T77528		27	2
75160	T77613	5331-01-310-8449	7	6
06021	U13639	5330-01-257-9152	37	17
36479	X4-89	5310-00-350-4717	5	22
SA352	028062-0130		5	5
SA352	028062-5790		5	10
5T151	028063-0030	3110-01-425-9914	5	17
SA352	028065-0250		5	24
SA352	028099-3270	3020-01-470-4287	5	32
SA352	028099-3570	5977-01-470-4249	5	6
SA352	028100-2881	2920-01-361-2361	5	4
SA352	028200-2230	2920-01-360-7746	5	2
SA352	028300-3691		5	28
SA352	028306-0070		5	18
5T151	028327-0171	5340-01-425-9915	5	19
SA352	028371-1500	3020-01-435-8532	5	16
SA352	028501-6900		5	9
SA352	028510-0500		5	8
53867	0290800069		4	14
84760	10349		21	42
84760	11056		21	83
84760	11141		21	80
53867	1120591075		4	8
53867	1123429200	5306-01-470-3597	4	1
53867	1124303024		4	6
84760	11331		21	1
84760	11563		21	5
84760	11919		21	11
75160	12M7065	5310-01-423-5301	1	6
84760	12221		21	60

PART NUMBER INDEX

PART NUMBER INDEX

CAGEC	PART NUMBER	STOCK NUMBER	FIG.	ITEM
66836	19M7865	5306-01-470-1384	18	13
75755	19M7866	5306-01-326-4911	7	10
			10	6
			11	10
75755	19M7867	5306-01-330-8490	1	17
			3	5
			30	2
66836	19M7868	5306-01-470-5111	1	25
66836	19M7913	5306-01-470-5067	1	23
66836	19M7923	5306-01-470-1984	27	9
66836	19M7970	5306-01-470-1993	10	7
			11	12
			14	8
66836	19M7979	5306-01-470-4193	16	2
			29	13
66836	19M8039	5305-01-470-5089	1	19
66836	19M8062	5306-01-470-1994	13	2
66836	19M8163	5306-01-470-1960	13	5
66836	19M8291	5305-01-470-2417	29	5
66836	19M8292	5306-01-470-1372	31	7
66836	19M8306	5306-01-470-4194	27	11
66836	19M8317	5306-01-470-1389	29	6
66836	19M8553	5306-01-470-6008	11	5
84760	19834		21	33
84760	19837		21	102
84760	19844		21	103
84760	19873		21	77
84760	19894		21	78
84760	19895		21	105
84760	19906		21	101
84760	20127		21	90
84760	20128		21	89
84760	20129		21	88
84760	20224		21	55
84760	20240		21	16
84760	20243		21	24
84760	20355		21	25
84760	20359		21	3
84760	20475		21	8
84760	20803		21	99
84760	20956		21	9
84760	21146		21	30
84760	21147		21	36
84760	21194		21	41
84760	21213		21	87
84760	21709		21	73
84760	22109		21	14
84760	22134		21	6
84760	22238		21	35
84760	22601		21	19
84760	22649		21	32

PART NUMBER INDEX

CAGEC	PART NUMBER	STOCK NUMBER	FIG.	ITEM
84760	22988		21	98
84760	23056		21	38
84760	23107		21	40
84760	23197		21	27
0BDN4	23856		21	74
84760	23860		21	76
84760	23915		21	106
66836	24M7055	5310-01-470-0643	19	4
75160	24M7207	5310-01-255-2619	1	18
84760	24419		21	43
66836	26M4224	5315-01-470-4170	33	3
0BND4	26346		21	61
84760	26355		21	66
0BDN4	26358		21	70
84760	26359		21	65
84760	26361		21	72
84760	26427		21	4
84760	26470		21	68
84760	27244		21	13
84760	27245		21	92
84760	27599		21	22
84760	27601		21	94
84760	27602		21	34
84760	27603		21	39
84760	27606		21	20
			21	21
84760	27607		21	15
84760	27608		21	109
84760	27610		21	37
0BND4	28381		21	71
84760	28749		21	69
84760	28974		21	75
84760	28986		21	108
84760	29136		21	79
84760	29363		21	85
84760	29366		21	86
84760	29384		21	95
84760	29710		21	96
84760	30387		21	63
84760	30804		21	67
84760	31332		21	54
84760	32163		21	91
84760	32466		21	28
84760	32478		21	2
84760	32519		21	29
84760	32533		21	31
84760	32548		21	104
84760	32636		21	110
84760	32642		21	107
84760	32787		21	62
84760	32859		21	97

PART NUMBER INDEX

CAGEC	PART NUMBER	STOCK NUMBER	FIG.	ITEM
84760	32880		21	64
84760	32947		21	44
84760	32948		21	45
84760	33044		21	52
84760	33334		21	46
84760	33613		21	50
84760	33614		21	51
84760	33840		21	26
75160	34H283	5315-01-119-3116	31	10
75160	34H288	5315-01-197-3144	31	3
84760	34033		21	47
84760	34245		21	48
84760	34284		21	49
84760	34322		21	93
33968	36881696	4330-01-444-3729	10	4
66836	38H5154	4730-01-470-1288	7	7
75160	51M7040	5331-01-471-7464	20	14
66836	51M7042	5331-01-470-5159	7	8
75160	51M7043	5331-01-460-2668	13	3
66836	51M7044	5330-01-470-2010	29	4
SA352	90258-06001	5310-01-360-9083	5	13
39428	90592A028	5310-01-364-4303	14	3
SA352	90804-10050	5330-01-360-9028	5	15
53867	9121064071	2920-01-469-9250	4	5
53867	9121064255	2920-01-470-7383	4	9
Z0666	9121067023		4	15
53867	9122067026	5340-01-470-2539	4	10
53867	9122067056	6115-01-470-6402	4	2
53867	9122067305		4	11
53867	9123065806		4	12
53867	9123065832		4	3
53867	9191067042	6110-01-470-6410	4	13
SA352	949006-2620	5305-01-361-6742	5	21
SA352	949006-4730	5305-01-470-1585	5	30
SA352	949007-6530	5305-01-470-1599	5	11
SA352	949011-5700	5310-01-360-1690	5	14
SA352	949016-0220	5310-01-361-2649	5	20
SA352	949042-2270	5306-01-361-5612	5	12
SA352	949100-2160	3110-00-100-2357	5	3
SA352	949100-2210	3110-00-100-2364	5	1
SA352	949100-2590	3110-01-470-3870	5	3
SA352	949100-5280	3110-01-470-3864	5	3
5T151	949120-0240	3110-01-361-0251	5	27
0E5W9	949170-4070	5360-01-381-6121	5	26
0E5W9	949173-0571	5360-01-374-8301	5	7
53867	9900069006	3110-01-470-3616	4	4
53867	9900069008	3110-01-470-3618	4	7

By Order of the Secretary of the Army:

ERIC K. SHINSEKI
General, United States Army
Chief of Staff

Official:

JOEL B. HUDSON
Administrative Assistant to the
Secretary of the Army
0029302

By Order of the Secretary of the Marine Corps:

RANDALL P. SHOCKEY
Director, Program Support
Marine Corps Systems Command

By Order of the Secretary of the Air Force:

MICHAEL E. RYAN
G eneral, United States Air Force
Chief of Staff

Official:

LESTER L. LYLES
General, United States Air Force
Commander, AFMC

DISTRIBUTION:

To be distributed in accordance with the initial distribution number (IDN) 256645 requirements for TM 9-2815-259-24P.

www.ingramcontent.com/pod-product-compliance
Lightning Source LLC
Chambersburg PA
CBHW080422030426
42335CB00020B/2546